THE
FOLKLORE
CALENDAR

THE LONGPARISH MUMMERS.

Frontispiece.

THE FOLKLORE CALENDAR

GEORGE LONG

SENATE

The Folklore Calendar

First published in 1930 by Philip Allan, London

Copyright © George Long 1930

This edition first published in 1996 by Senate,
an imprint of Random House UK Ltd,
Random House, 20 Vauxhall Bridge Road,
London SW1V 2SA

All rights reserved. No part of this publication may be reproduced, stored in a retrieval system or transmitted, in any form or by any means, electronic, mechanical, photocopying, recording or otherwise, without the prior permission of the copyright owners.

ISBN 1 85958 040 8

Printed and bound in Guernsey by
The Guernsey Press Co Ltd

CONTENTS

PREFACE	. . . .	1
JANUARY	New Year's Day – Up-Helly-Aa at Lerwick, Shetland Isles – Wassailing the Apple Trees .	7
FEBRUARY	Benediction of the Throat – Shrove Tuesday – Street Football on Shrove Tuesday – St. Valentine's Day	17
MARCH	The Tichborne Dole, and some other Mediæval Charities	27
APRIL	Hocktide at Hungerford – Easter Customs – All Fools' Day – Vigil of St. Mark . . .	41
MAY	May Day Ceremonies – The Furry Dance, Helston, Cornwall – Well-Dressing Ceremonies in Derbyshire – Well-Dressing and Holy Wells – The Morris Dancers – The Festival of the Dunmow Flitch – Whit-Monday at South Harting, Sussex – The Knutsford Festival .	65
JUNE	Riding the Marches at Hawick, Scotland – Town Criers' Contest at Pewsey, Wiltshire – Midsummer Morning at Stonehenge – The Druids' Festival at Stonehenge . . .	123
JULY	The Historic Tynwald Ceremony – The Vintners' Company	147
AUGUST	Punch and Judy – Wardmote of the Woodmen of Arden, Meriden, Warwickshire – The Lore of Harvest – Rush-Bearing at Grasmere – Highland Games and Gatherings . . .	153
SEPTEMBER	The Dance of the Deermen – Some Old English Country Fairs	169
OCTOBER	Fox-Hunting, Etc. – Hallowe'en . . .	195
NOVEMBER	Bonfire Night – London's Lord Mayor's Show, and Mayoral Customs Generally – Wroth Silver : St. Martin's Day	203
DECEMBER	The Christmas Ceremony at Killin, Perthshire – The Boy Bishop of Berden, Essex – The Christmas Mummers	213

ILLUSTRATIONS

The Longparish Mummers		*frontispiece*
The Norse Galley	*facing page*	10
The Tichborne Dole	,,	28
The Wayfarer's Dole at St. Cross, Winchester	,,	38
The Town Crier, Wells	,,	38
Hocktide at Hungerford	,,	46
Shoeing the Colt	,,	48
A Biddenden Cake	,,	56
The King's Maundy Ceremony	,,	56
May Day Ceremonies	,,	70
Holy Wells: Tissington	,,	82
Holy Wells: Tissington and Glastonbury	,,	86
Holy Wells: Harbledown and the Well of St. Keyne	,,	92
Mr. Wells and his Troupe of Morris Men	,,	102
The Morris Dancers	,,	106
Mr. Wells conducts his Troupe	,,	112
Midsummer Morning at Stonehenge	,,	132
The Procession from the Hele Stone	,,	138
The Entrance at Dawn	,,	138
Entering the Temple	,,	144
The Procession round the Slaughter Stone	,,	144
Mary Lee of the New Forest	,,	182
Gipsies at a Fair	,,	194
Romany Children	,,	194
Bonfire Day at Lewes	,,	206
The Overton Mummers	,,	218
The Longparish Mummers	,,	222
The Longparish Mummers	,,	226
The Overton Mummers	,,	232

PREFACE

THE object of this work is twofold. Firstly to interest modern readers in the many picturesque old customs which still remain to us, and secondly by such interest, to preserve from extinction some which are dying out. This can be done by publicity. I have already given new life to some very interesting ancient usages by a liberal 'write-up' in the Press, which had the result of attracting Press photographers to them next year. It is hoped that this book, which for the first time endeavours to provide a Calendar of the Folklore Year, may still further quicken interest in many delightful survivals of the past that are in some cases simply dying out through lack of public interest.

There is, however, a further reason why these old usages should be preserved, since the study of their remote origins provides valuable knowledge of the very beginnings of human beliefs. Just as the scientist can construct a ' missing link ' from a few fragments of skull and bone ; or picture to us some vast and terrifying monster of the Liassic or Jurassic epoch from some scanty vestiges imbedded in the rocks ; so can the student of folklore discover in these ancient customs ' fossil remains ' of human beliefs. From these scanty details we can form some idea of that primæval Tree and Nature Worship, which may well have been the Mother of All Cults, and traces of which can be found in the ritual of every religion.

The almost modern study of Comparative Religion has already taught us how many of them contain elements of similarity. But it is not my purpose to weary the reader with theological speculations ; it is rather to take him from the close atmosphere of the study to the Great Outdoors, and show him some of the quaint and beautiful old customs which are still carried on in our matchless countryside. We shall see the Mummers perform their eight-hundred-year-old folk-drama in front of a picturesque sixteenth-century cottage ; we shall watch the Morris Men capering on green lawns in a lovely old-world garden in all the glory of an English May Day. We shall travel to Stonehenge on Midsummer Morning and wait in the purple dusk till the Orb of Day rises over the Stone of Doom, where once the victim lay bound and helpless till his death shriek was drowned in the blare of trumpets which welcomed the coming of the Sun-god. We shall see the children dancing round the Maypole, and enjoy the Coronation of the May Queen of London, and the homage paid to her by all the lesser May Queens.

We shall follow the Tutti-men on their rounds, and make sure that the maidens of Hungerford permit their chaste salute, or pay the fine instead. We shall join in the noisy throng who hurry to Lewes, and see its vivid and delightful survival from the seventeenth century. We shall discover many homes of ancient peace, where old-world costumes are still worn ; or where age-old hospitality is dispensed with picturesque ceremony. We shall see the Well-Dressing at Tissington, and learn something of the lore of holy well and magic pool. We shall visit some ancient Fairs, and learn from the Romany Folk who still frequent them something of the romantic history of that fast vanishing race.

This work is the fruits of twenty-five years' study of the subject, and the scenes described have been personally visited. The subject of Folklore is, of course, of world-wide interest and application, but covers a field so vast that I have deliberately restricted the scope of this book to our own country. There can be no doubt, however, that the writer who has travelled afar is best fitted to tackle a subject of this kind ; and wanderings in many lands and in four out of the five continents have helped me to take a wider view of the subject than would otherwise be possible.

In all of them I have seen traces of those primitive beliefs to which I refer, not only among the classic ruins of Greece and Rome, but in tropic forests and among the teeming millions of the East. Wherever we go, from Arctic wastes to tropic forests, in the Old World or the New, we find everywhere the most interesting of all subjects to the thinker—the Genus Homo—Crown of Creation, and masterpiece of the great Architect of the Universe.

> ' Know then thyself, presume not God to scan,
> The proper study of mankind is man.'

While the material in this book has been obtained mainly by visiting the places described, and collecting information on the spot, no man could accomplish a work of this extent without some help from others. I desire to tender my grateful thanks to the many friends of all classes in many parts of the country who have assisted me with information both orally and by letter. I specially desire to mention that veteran archæologist Sir William W. Portal, Bt., M.A., F.S.A., of Laverstoke, at whose house I first saw the Mummers, Sir Joseph Tichborne, Bt., for permission to witness the Dole on several occasions, and Fr. Lion, also of Tichborne, for information and access to documents, etc., Mr. Lawrence E. Tanner of the Royal Almonry for permission to photograph and witness the King's Maundy Ceremony, the Vice-President of Magdalen College for permission to ascend the college tower during the May Day ceremony; the Rev. R. H. Harvey of Barwick-in-Elmet Rectory for information on local May Day customs. I have also to express my gratitude to Mr. Henry A. Benyon of Ufton Court for permission to view the Ufton Dole; Mr. Sam Bennett of Ilmington and Mr. W. Wells of Bampton for help and information about the Morris Dancers; the leaders of the Mummers in both the Longparish and Overton Troupes for much assistance in obtaining photographs and the oral text of their play, which they say had never been printed before I had it.

In addition to this I have, of course, derived much information from writers, living and dead. I need hardly mention *The Golden Bough*, that Bible of folklore students. I have quoted from Chambers's *Book of Days* and Brand's *Popular Antiquities*, as every writer on Folklore must ; I have also derived items of interest from Mr. Henry Walker about Knutsford ; and from Mr. Edmund Barber (of *Country Life*) about the Mummers.

I have derived some information about Hawick from Mr. S. R. Crockett's *Scott Country*, and I have quoted a paragraph or so from Mr. William Radcliffe's book on the Isle of Man.

I have taken a short paragraph on Wroth Silver from Mr. J. Charles Cox, LL.D., F.S.A., in his work on *Warwickshire*. I have also obtained the information on Rush-Bearing at Killin from Dr. George C. Williamson's most interesting work, *Curious Survivals*. I have also included a brief reference to the Dance of the Deermen from Mr. Charles Masefield's work on *Staffordshire*.

One difficulty in preparing a work of this kind is that dates are liable to change, and some have actually done so in the period taken to prepare this book. I am therefore especially grateful to Mr. L. A. de Meredith, O.B.E., for permission to bring them right up to date by verifying them from the Calendar of the Travel Association. He and his fellow-workers in this excellent association (which aims at attracting visitors to Britain from abroad), have gone to much trouble in checking these dates during the present year

(1930). It may be taken therefore that they are all correct for the moment, but readers can do much to assist the publishers in keeping future editions up to date by notifying them of changes which may take place.

JANUARY

NEW YEAR'S DAY
UP-HELLY-AA AT LERWICK, SHETLAND ISLES
WASSAILING THE APPLE TREES

NEW YEAR'S DAY

MR. G. K. CHESTERTON once remarked that even if Christmastide and New Year disappeared from our calendars, and their traditions vanished from human memory, the Winter Solstice would retain its magic.

This is due to the fact that in our own latitude, and indeed over a considerable area of the world, New Year's Day is the happy moment of the turn of the year when the sun climbs daily higher in the heavens, the days grow lighter and longer, and the whole earth gathers itself together for the renewal of Spring.

Our method of calculation is, of course, derived from the Romans. The ancient Jewish calendar opened with the 25th of March, which as the commencement of Spring is a still more obvious period for the birth of the New Year, though January is astronomically more accurate. The name is derived from Janus, the two-faced Roman god, since the first month of the year might be regarded as looking both ways, back to the old year and on to the new.

There are still a few old customs relating to New Year's Day, particularly in the North of England and Scotland. Scots (as all Londoners who venture near St. Paul's about midnight on New Year's Eve are well aware) have a particular fondness for the festival of New Year.

In these northern parts of our island the quaint

old custom of 'First Foot' is still observed as the clock chimes in the New Year. A tall dark man is required to bring luck to the household by first crossing the threshold; and those of the correct type are greatly in request by their friends to assure them a prosperous New Year. He knocks at the door, and when it is opened, pushes through and proceeds to the domestic hearth without a word of apology or explanation. Arrived by the dying embers of the fire, he places new fuel thereon, which must have been brought in by him; and in some districts salt and bread are placed on the table in the same solemn silence.

The visitor must then be offered wine and cake, after partaking of which he wishes the household good luck in the days to come; and in bygone days it was religiously believed that this quaint ceremony would really ensure benefit. Modern education, however, has lessened faith in this and many other old beliefs, and it is now not uncommon for parties of frolicsome young folk to deputise in the 'wee sma' hours' for the tall dark man, who is the traditional harbinger of good.

January 23rd

UP-HELLY-AA AT LERWICK, SHETLAND ISLES

THE celebrated Carnival of Up-Helly-aa at Lerwick, Shetland, is one of the most interesting and picturesque survivals in all Britain, while the burning of the Norse galley is one of the most impressive things of its kind that can be seen to-day. Though on a smaller scale than Bonfire Night at Lewes, it surpasses this in picturesqueness of detail and antiquity, since the burning of the Norse galley undoubtedly goes back to the funeral rites of those hardy Viking chiefs whose corpses were set adrift in a blazing galley *en route* for Valhalla.

A full description of the latest celebration appeared in the *Shetland News* of January 30th, 1930.

The celebrations consist mainly of a great torchlight procession to Norse music, with magnificent figures of Vikings in full armour escorting the galley, named the *Fragæth*. The vessel was $31\frac{1}{2}$ feet long and six beam, and was mounted on a rubber-tyred motor-chassis. The lofty dragon's head rose $8\frac{1}{2}$ feet above the platform, and the tail $7\frac{3}{4}$ feet, making about 12 feet in height from the ground. There were ten oars on each side, each painted blue with white blades, and fitted near the grip with a metal shield adorned with ravens; and a mast rose 15 feet above the platform.

Her captain was the Guizer Jarl (or head

THE NORSE GALLEY.
(By courtesy of the *Shetland News*.)

To face p. 10.

Guizer) with his squad of noble Vikings in their impressive costumes, and great winged Norse helmets. The vessel was dragged through the streets by Vikings, and escorted by others carrying torches, while a double line of flame followed behind. The Guizer Jarl was an imposing figure in his glittering coat of mail and lofty winged helmet, as he stood proudly at the dragon-head of his galley. Proceedings commenced at 7 p.m. by firing a rocket, and the procession started at 8.30 headed by two bands, and followed by 320 torch-bearers.

The 1930 celebration was perhaps the finest yet seen, as not only was the weather very favourable, but the attendance was exceedingly good, and the whole town of Lerwick seemed steeped for the time in the old Norse traditions. There is Viking blood in the Shetland folk to-day, and these hardy islanders are worthy descendants of the noble Vikings, who so fearlessly faced the perils of the deep. There is nothing in British folklore to equal in impressiveness this wonderful procession, as seen in the ruddy glare of the torches. At its head goes the mighty dragon-headed war galley, commanded by the splendid Guizer Jarl in his panoply of war. Behind him are the ranks of Vikings, the torch-light flashing on their coats of mail and glittering on the broad shields emblazoned with the raven of their ancestors. They are splendid in appearance, and some are still more magnificent perhaps in fact, as those can say who have seen them in their frail fishing craft during the gales of winter. The fisherman risks his life to get the fish, and then, if he is paid at all (and sometimes he is not), his remuneration is a disgraceful fraction of the amount the pampered fishmonger receives for merely handing it over his counter. Mr. John Grierson (who filmed the fishing fleet for *Drifters*) pointed out in a recent article that billions of fish were lost to the public in 1929, and the fishermen were prepared to deliver them to the markets at 12/- a 1,000, but the retailers' ring refused to have them, because herrings must never be sold at less than 2*d*. each. In one case a quarter of a million fish were thrown into the sea by the ruined and broken-hearted fishermen, because the grasping retailers would not have them and would rather waste good food than sell it cheap now and again.

It is just as bad on other coasts. A splendid old Devonshire fisherman told me with a sigh that he was lucky if he could make 2*d*. a lb. average on all his fish, and he added, " I never sees them less than 1/4 in the shop." Fishermen pray for a Mussolini, to send profiteering retailers out to sea with the fishing fleet – dirty weather essential – so that each should learn for himself what it costs in human effort, courage and endurance, aye, and precious lives as well, to win from the sea those fish which are so lightly esteemed by the fishmonger when he buys, but so differently valued when he sells. And here in these bleak northern islands, so far north that midsummer sunset glow mingles with early dawn, here in a latitude north of Cape Farewell in Greenland we have a hardy population of fishermen and crofters. No place in the islands is more than three miles from the sea, and in stormy weather it seems impossible ever to get away from the roar of the breakers. These splendid men are worthy successors of the Vikings who once owned the islands. Actually they are only British because the Danes never paid James III the promised dowry of his wife, Margaret of Denmark. They handed over Orkney and Shetland as a pledge till the money was paid. That was in 1469, and the cash is still outstanding.

As we watch the majestic galley slowly advance, flanked by a double line of torches, and manned by the splendid figures of its Viking crew, we are tremendously impressed by the wonderful sight and the memories it awakes. Slowly, solemnly it proceeds, headed by an imposing squad (No. 1 Squad of Guizers) in Viking dress, while the two bands play Norse tunes, and anon ' John Brown's Body,' more familiar to Saxon ears. Through streets whose names recall the past – King Harald Street and King Eric Street – it passed into the Commercial Road, entered the Lower Esplanade, and moved along the sea-front. Now maroons were fired from the Fort, and rockets from ships in harbour, and 'mid the crashing and flashing of the fireworks, the galley came to rest. The torch-bearers formed a flaming circle around, and the Guizer Jarl – a splendid figure in the stern – called for 'The Norseman's Home.' This was sung in a most impressive manner; then the Norsemen left the ship, a bugle call sounded, and 320 torches were flung into the doomed ship. Within a few minutes she was a mass of flames, and the dragon-head and mighty scaly tail crashed into the cauldron of fire below. As the splendid outlines of the noble vessel stood out stark against the blaze, one was irresistibly reminded of the hundreds of gallant Vikings who made their last voyage thus outlined in flames, and so went to Valhalla with the brave ship they loved so well.

WASSAILING THE APPLE TREES

AT Carhampton near Minehead, Somerset, the old custom of Wassailing the Apple Trees is still observed on January 17th.

The men of the village form wassail parties, and go to certain orchards, where the farmer and his men join them. Then, standing round the trees, they sing the old wassail song, the last verse of which goes thus :

'Old apple tree, old apple tree,
We've come to wassail thee,
To bear, and to bow apples anow,
Hats full, caps full, three bushel bags full,
Barn floors full and a little heap under the stairs.'

With due ceremony a piece of toast is placed in the fork of the biggest tree, ostensibly for the robins ; but really the old folk say to propitiate spirits who watch over the apple trees.

This custom was formerly observed also in eastern Cornwall, when cider was drunk in the orchard and the remains of the liquor thrown at the trees. The rite was supposed to ensure a good crop and is no doubt a survival of a very ancient practice of sacrificing to the spirit of the harvest.

FEBRUARY

BENEDICTION OF THE THROAT
SHROVE TUESDAY
STREET FOOTBALL ON SHROVE
  TUESDAY
ST. VALENTINE'S DAY

February 2nd

BENEDICTION OF THE THROAT, ST. ETHELDRED'S CHURCH, ELY PLACE, HOLBORN, LONDON, W.C.

FEBRUARY 2nd is the Feast of St. Blaise, Patron Saint of Wool-combers, who is said to have been martyred in the year 316 by sharp iron combs, which tore his flesh to pieces. Tradition has it that, as St. Blaise was going to his death, he saved the life of a boy who was choking to death from a fish-bone in his throat. One touch of the Bishop's fingers (he was an Armenian prelate) on the lad's throat dislodged the bone, and the boy was saved.

Every year on this day, then, a quaint ceremony is performed in this ancient Catholic church. Two long candles are blessed, dedicated to St. Blaise, and tied together in the form of a St. Andrew's Cross with a piece of ribbon. Persons suffering from throat trouble then approach and kneel, and the priest will touch their throats with the now lighted candles, so that the crosspiece is below their chins, and the flames, of course, cannot touch them. As he applies the candles he says, " May the Lord deliver you from the evil of the throat, and from every other evil."

The services are held at 1.15 p.m., 6 p.m., 8 p.m.

SHROVE TUESDAY

THIS is well known to country folk as Pancake Day. In mediæval times the day was devoted to feasting, doubtless as a set-off to the dismal period of fasting which followed. It is supposed that pancakes were made on this day in order to use up the household stores of fats or butter, which before the Reformation were strictly forbidden to the people for the period of Lent. There are several interesting customs relating to pancakes which are still observed. One of these is the Pancake Bell, a survival of the pre-Reformation practice of calling the people to church to confess their sins on Shrove Tuesday, by ringing a bell. The name Shrove Tuesday, of course, refers to the fact that after confession the faithful were shriven.

At Olney, Bucks, this Shrove Tuesday bell is still rung, though to-day for a much more cheerful reason. The first chimes call on the inhabitants to mix up the materials for the pancakes, and then a big bell tolls as a signal to eat them. Immediately girls rush to the church with frying-pans in their hands and the pancakes merrily sizzling within. The reason for the race is to have the honour of being the first to serve pancakes to the ringers. There is another interesting Pancake Scramble at Westminster School on Shrove Tuesday, when twenty boys – one selected from each form – line up for the Tossing the Pancake ceremony. The Dean gives a guinea to the boy who secures the pancake, or the largest fragment thereof, and the cook takes pride in throwing it over the beam in the lofty roof.

I find a reference to this custom in the *Gentleman's Magazine* for 1790. The pancake was then thrown over a bar which divides the upper from the lower school, and apparently everybody scrambled for it.

STREET FOOTBALL ON SHROVE TUESDAY

IN mediæval times Shrove Tuesday was not only a day of feasting, but frequently a kind of mild saturnalia. At any rate, much licence was permitted to the prentices, and the remarkable old custom of Street Football is undoubtedly a survival of the rough sports of the prentice boys.

There are three places in England where this old custom is still observed, Ashbourne (Derbyshire), Atherstone (Warwickshire) and Chester-le-Street (Durham). In each of these places the old observance is maintained with considerable vigour, and a large number of men and boys take part in some rather rough sport. At Chester-le-Street the tradesmen are obliged to protect their shop windows with wooden hoardings, or they would certainly be smashed by the rough element of this unlovely town. The old rhyme says :

' Chester-le-Street has a bonny, bonny church,
With a broach[1] upon the steeple,
But Chester-le-Street is a dirty, dirty town,
And mair shame for the people.'

[1] Spire

February 14th

ST. VALENTINE'S DAY

> ' Muse, bid the morn awake,
> Sad winter now declines,
> Each bird doth choose a mate,
> This day St. Valentine's :
> For that good Bishop's sake
> Get up, and let us see
> What beauty it shall be
> That fortune us assigns.'
>
> *(Drayton.)*

THERE were at least two ' Saint ' Valentines, and quite a number of martyrs, for the name was almost as common in ancient Rome as Smith with us, or Durand in France. However, the term St. Valentine is generally applied to a priest who was put in chains by the Emperor Claudius for helping Christian martyrs. While he was in prison he preached with so much power that his gaoler, Asterius, was converted and baptised by him. This further enraged the Emperor, who caused him to be cruelly beaten with clubs and finally beheaded on the Flaminian Way (February 14th, A.D. 269).

Doubtless this is the individual to whom the Church has dedicated February 14th, but there is no connection whatever between any of the saints and martyrs named Valentine and the old customs associated with the day – save only the accident of the date. The early Church, following its general policy of Christianising pagan customs rather than abolishing them, substituted the name of a saint for a pagan deity, prohibiting the more grossly sensual observances of paganism. As the festival of the Lupercalia formerly commenced about the middle of February, it was very natural to rechristen this with the name of a saint who was martyred at the same period. The Lupercalia were dedicated to Pan and Juno, and one of the ceremonies consisted in placing the names of young women into a box from which they were drawn as chance directed. This seems clearly to indicate the origin of the mutual love missives which have become the main feature of the day. Unfortunately the observance is declining, and will probably ultimately disappear, since the fair sex have now obtained so much liberty in choosing their male friends that no Valentine's Day is needed. Mr. Pepys enters in his famous *Diary* on Valentine's Day, 1667 :

'This morning came up to my wife's bedside (I being up dressing myself) little Will Mercer to be her Valentine, and brought her name written upon blue paper in gold letters, done by himself, very pretty and we were both well pleased with it. But I am also this year my wife's Valentine, and it will cost me £5. [Here comes a characteristic aside.] But that I must have laid out if we had not been Valentines.'

A well-known and picturesque idea connected with St. Valentine's Day is that on this day the birds choose their mates. Robert Herrick (that merry seventeenth-century clergyman who gave so much good advice on the lines of ' Gather ye rosebuds while ye may,' but never himself got married at all) reminds us of this pretty idea in these words :

' Oft have I heard both youths and virgins say,
  Birds chuse their mates, and couple too, this day,
  But by their flight I never can divine
  When I shall couple with my Valentine.'

In many countries, with but slight variations, the same ceremonies were carried out on the Eve and the Day sacred to St. Valentine. The commonest was the placing of an equal number of boys' and girls' names on pieces of folded paper, which were placed in separate vessels and drawn in pairs. This was supposed to indicate which couples would ultimately mate.

A writer in the *Gentleman's Magazine*, published in 1779, describes a sport among the children in Kent, in which the girls burned in triumph a figure which they had stolen from the boys, called the Holly Boy, while the boys were doing the same thing with another figure called the Ivy Girl. Then followed a sham fight for each other's trophies. Several methods of divining a future lover were employed by the girls, the following being very drastic. On St. Valentine's Eve five bay leaves were pinned on the damsel's pillow, one at each corner and one in the middle. This was bad enough, but worse was to follow. The yolk was removed from a hard-boiled egg and replaced by salt, and then the whole, eggshell and all, was eaten. And so to bed and dream of the future husband.

Another and pleasanter form of divination was to write the names of several potential lovers on pieces of paper, roll them up in balls of wet clay, and put them in a bowl of water. As the clay dissolved, the paper would come to the surface, and the first up would be the lucky man.

In my youth the stationers' shops were full of gross caricatures of the most offensively personal character, known as Valentines. These criminal libels consisted of a most insulting portrait of some individual together with a quatrain of feeble verse calling attention to some physical or moral characteristic of the person, so worded as to be likely to lead to murder if he found out who sent it. These lampoons were forwarded under a cloak of the discreetest anonymity, and were regarded in certain social circles as delightfully humorous. The milkman had the grossest references to the alleged malpractices of his trade with the pump, the grocer was reminded of his fondness for sanding sugar, while the old maid was assured of the utter hopelessness of her unending quest. Happily the spread of education and the improved manners of the people have almost extinguished these gross pleasantries, and such Valentines as survive to-day are usually sugary effusions of the 'hearts and love knots' variety.

MARCH

THE TICHBORNE DOLE, AND SOME OTHER MEDIÆVAL CHARITIES

March 25th (yearly)

THE TICHBORNE DOLE

THERE are few more picturesque survivals in England to-day than the ceremony of the Tichborne Dole, and none that can exceed it in dramatic and historic interest. There are few indeed among the ancient families of England who can trace their descent back to Saxon times, or who have lived on their ancestral acres a thousand years and still remain to-day. When we add to this an amazing legend in the past, a picturesque custom in the present, and one of the most sensational trials in history, it will be seen that the House of Tichborne has claims to a large share of public interest. The present Baronet is Sir Joseph Tichborne, and he is still living on the very spot where his ancestors were residing before Duke William came a-conquering in 1066.

The place is one of amazing interest, both for romantic legends of the past and sensational happenings in the present. As recently as the winter of 1926 an old lady died in South Stoneham Workhouse, Hants. Though known to the authorities as Mrs. Orton, she insisted on being addressed as Lady Tichborne, being the widow of that amazing adventurer the Tichborne Claimant, to whom I shall presently refer.

Before describing the remarkable ceremony of

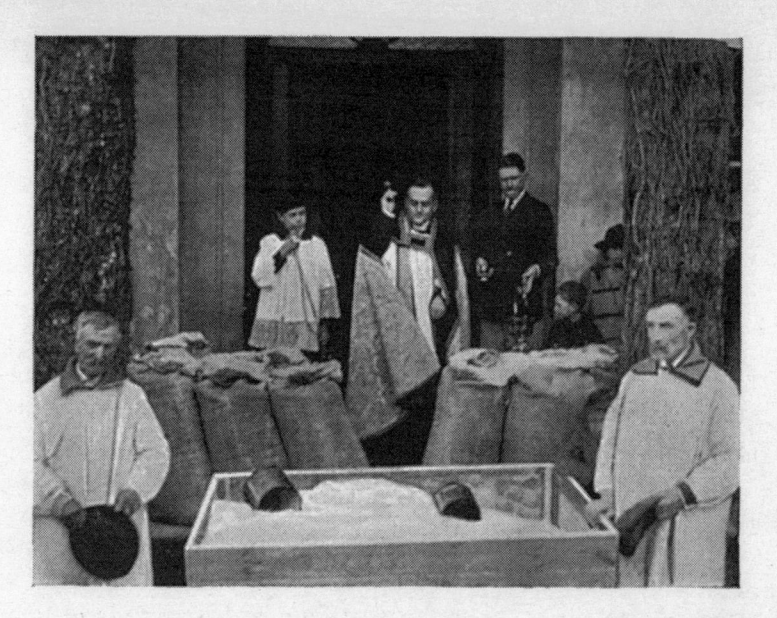

BLESSING THE FLOUR.

SPRINKLING THE FLOUR WITH HOLY WATER.
THE TICHBORNE DOLE.
In each photo, though taken in different years, the lad dressed as an Acolyte is the youthful Tichborne heir.

To face p. 28.

the Tichborne Dole I will relate the events which are said to have led to its foundation.

Away back in the twelfth century, when Henry I was King of England, a certain Sir Roger Tichborne ruled the estate with a rod of iron. His wife was the gentle Lady Mabella, who had come to him, richly dowered, from Limerston, near Brighstone, Isle of Wight. She was known and loved throughout the whole countryside for her pious and charitable disposition, the very opposite of his own. After a long life, well filled with good deeds, she lay at last on her deathbed. She had no fear of death, but her gentle soul was troubled by the thought of the distress which would fall on the poor of the village when they could no longer come to her for help. In those days the condition of the poor was dreadful. There was neither 'Dole' nor poor relief, and many of the wretched inhabitants actually died in their miserable mud and wattle huts when harvests failed. So she pleaded with her husband to set aside a piece of land large enough to provide a dole of bread to all comers on the day of the Annunciation. She reminds us of the gentle Lady Godiva, who also pleaded the cause of the poor ; and Sir Roger's answer to it was quite in the best style of Earl Leofric at Coventry. Each of them undertook to grant his lady's desire, if she, on her part, would undertake something which they believed to be impossible. Sir Roger did not suggest an equestrian performance for the Lady Mabella, but instead something which must have seemed equally impossible for a dying woman. Seizing a blazing stick from the open hearth, he held it aloft and harshly declared that he would give all the land she could walk around while the flame was burning. The sick woman closed her eyes for a moment in prayer for Divine help, then calling her women, ordered them to carry her outside. They placed her gently on the ground, and she essayed to stand, but could not.

Then falling on her kness, she crawled – and crawled as never woman crawled before. The knight, already repenting his rash promise, followed her outside, still holding the blazing brand. But the wind seemed to have dropped, and the flame burned clear and steady without a flicker. Before half was consumed the crawler was a mere speck in the distance, and had already turned back towards her starting-point. Breathlessly the servants watched the fatal flame as it slowly ran its course, and a race as for very life proceeded between Lady Mabella and the expiring brand. But before the blaze gave a final spurt and flickered out, she had returned to her starting-point, and twenty-three acres had been encircled. The serving-maids carried her gently to her pallet, and Sir Roger followed – savagely flinging the charred stump into the fire. As he approached Lady Mabella beckoned with a thin finger. " Listen to me again, my husband, for my time is short. God has heard my prayer, and the land thou hast given shall provide a Dole of Food for my poor, and the day I appoint shall be that of our Lady's Annunciation. And"—here her voice grew stern with warning—"let no man dare break this solemn promise, or tamper with so sacred a gift, for then a curse shall fall upon him, and upon his house. The fortunes of the family shall fail, the name of Tichborne shall be changed, and the family shall die out. And as a sign thereof there shall be born a generation of seven sons, followed by one of seven daughters."

She fell back on the pillow. The maids bent over her anxiously and then burst into violent weeping. Lady Mabella was dead.

Eight hundred years have passed, and the Tichbornes still reign over their ancestral acres. The twenty-three acres still produce their Dole of Food for the poor, and are known as 'The Crawls' to this day; and on each Lady Day the flour is given to the poor. No Tichborne will risk that terrible curse which once was almost fulfilled.

In 1796 Sir Henry Tichborne stopped the distribution of loaves of bread, and instead gave money to the Church. Then a remarkable thing happened. I have seen the actual family-tree of the Tichbornes, and there—plainly marked—is the generation of seven sons followed by seven daughters, as prophesied. The Dole was promptly restored to its original form, and the curse was staved off, but more trouble for the family was brewing.

In January 1852, Roger—the then heir—after

a disappointment in love, sailed in the *Bella* for Valparaiso. The vessel was never heard of again, and she was presumed lost; fourteen years later his younger brother Alfred succeeded to the title and estate.

But a mother's heart can never forget, and though all others believed him drowned, Lady Tichborne refused to do so; and for years she continued to advertise in the home and colonial newspapers for news of her lost son. Finally news was received from him (?) in Australia, and the 'Claimant' (as he was afterwards styled by the world's Press) was invited to return to England, and money was provided for the voyage by Lady Tichborne. The poor old lady received him gladly and accepted him as her lost boy; but the trustees and legal advisers of the estate did not share her opinion and refused to concede his claims; although he was certainly in possession of a remarkable store of knowledge of intimate facts in Roger Tichborne's life, and bore a resemblance to him.

Thereupon the 'Claimant' brought an action to gain possession of the estate, and so began the first of the two celebrated Tichborne Trials. The trial lasted for four months, but he broke down under cross-examination, particularly when he was pressed about his 'Varsity life. It is said that when asked if he remembered the 'Bridge of Asses' (i.e. the Pons Asinorum, the famous proposition of Euclid), he replied, "Yes, I have often fished from it!"

After the fiasco of the first trial, he was committed to prison on a charge of perjury, and after another hearing, which lasted from May 1873 to February 1874, he was sentenced to 14 years' penal servitude. In spite of the verdict of the two trials, thousands of people have always believed that Arthur Orton really was Roger Tichborne. "How else," they declared, "can we explain why people who had known Roger Tichborne recognised him in the Claimant? And how could Arthur Orton, a Wapping butcher, have come into possession of so many private details in the life of a Hampshire landowner he had never seen?"

The case was very mysterious, but I have never believed that Orton was any other than a fraud. When at Tichborne House, Sir Joseph kindly showed me a number of the Claimant's letters, and it seems incredible that such illiterate stuff could have come from the pen of an educated gentleman who had spent several years at the University. All his 'I's' are written with a small 'i,' and his spelling was beyond belief. Torquay, for instance, was rendered 'Torkey.' But a solution of the puzzle has recently come out. An unsigned article in the *Cornhill* (July 1929) gives a romantic story, which fully explains the mystery.

According to this writer, Roger Tichborne sailed from Valparaiso in the *Bella* in 1852, the vessel was sunk, and Roger was picked up by a ship, the sole survivor. He had been clinging to a bit of wreck till his mind was gone, and was taken to Australia as a hopeless imbecile. As he was quite harmless, however, they soon let him out, and he wandered about the country doing odd jobs. Here he met Orton, then working as a butcher in Australia, under the name of Castro; but who *was really his illegitimate half-brother and very like him*. The poor idiot, though he did not know who he was, or how he got there, had a habit of maundering about his early home life; and Orton, knowing the family, soon suspected who he was. When, eight years after Roger's disappearance, Orton saw Lady Tichborne's advertisement in an Australian paper, he went to the lawyer named – thinking to get a share of the reward. When shown into the office, Orton stated that he 'had come about the advertisement,' and the lawyer, not unnaturally, supposed that Orton was Tichborne himself, and addressed him as such. This gave the 'Claimant' his big idea, and won for him fame but not fortune.

This story may well be the secret of this dramatic romance of real life. I am told by those who knew him, that when Orton was released from prison, he was a slim silver-haired old gentleman of most aristocratic appearance. At the trial he had been very stout, and looked coarse rather than distinguished, and this together with his illiteracy is no doubt what caused him to lose the case. No one reading his letters – which I have seen – could possibly believe them to be written by a cultured gentleman who had been a 'Varsity graduate. But I must return to the Tichborne Dole.

The public are freely permitted to see the ceremony, which takes place every year on Lady Day (March 25th) with all the stately ritual of the Roman Church. My photograph gives a good idea of what occurs. A great bin, holding a ton and a half of flour, is placed in the porch, and behind it are grouped the principals. In the centre is the family Chaplain (the Rev. Father Lion) and on his left is Sir Joseph Tichborne, the present Baronet, with the youthful heir of the estate (dressed as an acolyte) on the priest's right.

The service, which is wholly in Latin, commences with prayers for the repose of the soul of the Founder (Lady Mabella), the flour is then sprinkled with holy water and blessed by the priest, and after some more prayers the flour is given away to the people of the village in the proportion of one gallon for each male in the family, and half a gallon for each woman and child.

So far as the legend is concerned, it may quite well be true, at any rate to the extent that the land which provides revenues for the Dole was a gift of Lady Mabella. I have ascertained that Limerston Manor, near Brighstone, Isle of Wight, came into possession of the Tichbornes somewhere about the period stated by the legend, and there is a deed executed early in the reign of

Edward I by Sir Roger de Tichborne, relating to a chapel at Limerston. This to some extent confirms the legend, since Lady Mabella was an heiress from the Isle of Wight, and her property would have passed to her husband on her marriage. I have also seen an actual charter signed by a Tichborne, relating to a property deal in the reign of King Stephen. These facts certainly confirm the antiquity of the family, and the Dole itself may well have originated in a gift by the pious Lady Mabella, since charity has always been a Christian virtue. It is also, of course, a fact that mediæval Doles were usually dated for the season of Lent – a period set aside by the Church from early times for prayer, fasting, and pious works acceptable to God.

Thus another pious lady founded the Ufton Dole, which is still distributed with old-world ceremony on the Friday after the third Sunday in Lent. This was Lady Marvin, who bequeathed a gift of 169 loaves to the poor of the district in the year 1583. In addition, nine persons receive clothing, consisting of five yards of flannel and eleven of calico each. Mr. Benyon, the present owner of Ufton Court, Berks, very kindly allows visitors to see this interesting ceremony on application. The distribution is made from a window at the back of the house, so permission to attend provides an opportunity of seeing the magnificent old house and the splendid gardens from the terrace. The house itself (not shown) is a fine example of Elizabethan architecture, with its nineteen picturesque gables, and splendid front which illustrates the characteristic E form of the period, said to have been adopted by patriotic builders as a compliment to 'Good Queen Bess.' The interior is even more interesting than the exterior, and contains no fewer than three cleverly constructed priests' holes, the most remarkable of which is at the south end of the long gallery. It is at the foot of a deep well-like hole, and is reached by a ladder.

Readers who wish to discover Ufton Court for themselves should note the following information. Take the Bath road from London, and after leaving Reading there are four roads to the left between Theale and Aldermaston, either of which will bring us to the Court, but the best route is to take the turning to the left close to the fiftieth milestone from London and proceed through Padworth. In this village, by the way, another Dole is given, which, like the Ufton Charity, is due to the generosity of Lady Marvin; and both have been carried on for nearly three and a half centuries, save for a gap in the wicked eighteenth century, when so many charitable bequests were misappropriated. Easter is specially favoured as a date for mediæval Doles. The most interesting of these is that of Biddenden (see page 56), and there is another at Ellington, a charming old-world village about five miles west of Huntingdon on the Thrapston road.

The distribution takes place each Easter Day

at the old village church, which is a landmark on the main road. The qualification for it is that each recipient must have slept in the parish overnight.

This was no doubt a charity to assist poor travellers, but was greatly limited in effect, owing to its taking place on one day in the year only. Before the Reformation the various monastic houses provided most generous hospitality for all wayfarers daily, and there were few parts of the country where a traveller on a main road could be more than a day's journey from such accommodation. There is only one such *ancient* charity still remaining in England to-day, and that is 'The Wayfarer's Dole' at St. Cross, near Winchester. The institution was founded about 1133 by Bishop Henry de Blois, and was at first an almshouse for thirteen poor men, and in addition gave free meals to a hundred poor folk daily. Half a century later the control of the Order was assumed by the Bishop of Winchester, and the number of poor persons fed daily was increased to 200; but unfortunately later Bishops appointed to the mastership young relatives — who stole its revenues and neglected the poor. The great William de Wykeham (founder of Winchester College, and virtual builder of the Cathedral as it is to-day) stopped this scandal; and his successor, Cardinal Beaufort, further endowed the charity by adding an almshouse of 'Noble Poverty' which was designed to give help to poor gentlefolk. Both charities are still

THE WAYFARER'S DOLE AT ST. CROSS, WINCHESTER.

THE TOWN CRIER, WELLS.

To face p. 38.

continued, and the two classes of inmates can be distinguished by the colour of their gowns. The poor gentlefolk (mainly decayed professional men) wear dark red, and the older foundation black; both carry the silver eight-pointed cross of the Knights Hospitallers.

The old men have delightful little houses, with a food allowance and 5/- a week pocket money. I am told that there is great competition to obtain places in the black-gowned section; but that sometimes there are vacancies among the red robes; happily professional men are seldom in need of the real help this noble institution affords.

But I am wandering from The Wayfarer's Dole, which is still given daily to all who knock at the Porter's Lodge. It consists of a piece of bread and a horn of beer, and is given without question to all callers up to a certain daily limit. Those who are really in need are given a much larger portion than the mere tourists who call from curiosity. And when you draw your ration, remember that this bounty has been provided daily for nearly nine hundred years!

The Porter who dispenses the charity wears a most picturesque costume; and the bread and beer are served on an ancient carved oak platter. The horns bear the arms of the order in silver and several are preserved in the Porter's Lodge which have been used by the great ones of the earth when visiting St. Cross.

There is a similar charity at Cowfold in Sussex, but it is of no historic interest, since the monastery was founded in 1877 only; and others can be found at other religious houses throughout the country, but of recent date.

APRIL

HOCKTIDE AT HUNGERFORD
EASTER CUSTOMS
ALL FOOLS' DAY
VIGIL OF ST. MARK

HOCKTIDE AT HUNGERFORD

THE old-world town of Hungerford is situated on the Bath road some nine miles to the west of Newbury. Few, however, of the legions of motorists who speed along that famous highway are aware of the remarkable celebrations which have been carried on there for many centuries and still continue. There is some dispute among the learned as to the exact significance of the word 'Hocktide,' but the idea that the festival originated to commemorate the massacre of the Danes on St. Brice's Day, 1002, is obviously wrong, since that sanguinary event took place on November 13th.

The fact, however, that the date changes with Easter (being actually the Monday and Tuesday in the week following Easter Monday) suggests that some pagan ceremony may have been tacked on to Easter. There is further evidence to suggest that it was a rent day, in fact at one time Hocktide and Michaelmas divided the year into a summer and winter half for landlords and tenants. There was a pre-Reformation custom called 'Binding Monday' which took place annually at Hocktide. On Monday the women levied toll on the men, by binding them with ropes till they paid a small fee to be released, or by stretching a rope across the highway and only lowering it to permit passage on payment. The next day (Tuesday) the men performed the same office for the women, and the remarkable festival at Hungerford is a survival of this. Thus in the records of the Parish of St. Laurence, Reading, under date 1499, we find these entries :

'Item, received of Hock Money gaderyd of women xx s. (20/-)
Item, received of Hock money gaderyd of men xiij s. (13/-)'

Those of my masculine readers who have suffered from feminine persuasion on flag days will note with interest that even in the fifteenth century the ladies were better at mendicancy.

In spite of this, the great day at Hungerford is now Tuesday, and it is the men who strive to extract money from the fair sex. Local tradition suggests that the festival was founded to commemorate the gift to the town by John of Gaunt (Duke of Lancaster) of certain manorial fishing and common rights. With it he gave an ancient horn which is carefully preserved in the Town Hall and was formerly used to open the proceedings. In 1634 a new horn was made for the purpose, and the old one carefully put away for safe custody. The 'new' horn bears this inscription :

'John a Gaun did give and grant the riall of fishing to Hungerford toune from Eldren stub to Irish Stil except some several mil pound.'

This horn is used to open the proceedings on Tuesday morning at 8 o'clock, and is blown from the balcony of the Corn Exchange by the town crier in full uniform. For some years it was the custom for the school children to assemble in the Broadway opposite the Corn Exchange when the horn was blown and sing this song by Mr. A. V. Murray.

> 'This ancient Market Place has heard
> For half a thousand years,
> The summons of a mighty horn,
> Time-honoured Lancaster's.
> The Tutti-men have trimmed their poles
> With Blossoms fresh and gay
> And kissed the merry bashful maids
> On Hocktide holiday.
>
> From Wolstone's vigilant retreat
> The White Horse hurries down
> And glares across where Inkpen's height
> Keeps watch above the town.
> Those rolling hills have heard the cry
> Of centuries of war,
> E'er Hingar's fatal ford received
> The Crescent and the Star.
>
> And sweet the twilight of the eve
> Steals down the Icknield Way;
> And Wayland's Cave is murm'rous still
> When night has closed the day.
> With voices of a sturdier past,
> Whispering the tales of old,
> Of Alfred and the fickle Earl
> And Stuart overbold.'

I am sorry to say I have not heard this song sung during the last few years, though it was well known in pre-war days, and at the Newbury Folk-Festival in March 1913 a party of Hungerford children sang it, and several other local songs, and received the highest award.

After the blowing of the horn, the bellman parades the town and makes the following announcement :

' Oyez, oyez, oyez. All ye commoners of the Borough and Manor of Hungerford are requested to attend your Court House at the Hall at nine o'clock this morning to answer your names, on penalty of being fined. God save the King.'

At 9 the Court of Feoffement sits. This is really an ancient manorial court, the twelve members of which (called Feoffees) are elected citizens, who act as Lords of the Manor in dealing with fishing and common rights. Persons claiming such rights have to attend and make good their case ; they also let part of the fishing to clubs etc., which are a valuable source of revenue to the town, and give the townsfolk an opportunity of sharing in their own amenities. This is a real advantage, as on some other rivers possessing good fishing, the old inhabitants complain that the monopolistic policy of the millionaires' trout-fishing clubs has actually deprived the people of their own rights. The wealthiest and most monopolistic of these clubs has actually adopted a policy of buying up all the little bits of second-rate fishing which were formerly enjoyed by local inhabitants, *not in order to fish them, but to prevent anybody else doing so.* I could point out some such stretches which formerly gave pleasure to local residents of moderate means, and are now closed to everybody, and used by nobody. They are not good enough for the millionaires to use, but too good for anybody else ! But I must return to my subject.

Meanwhile the two elected Tutti-men of the day attend at the Constable's house to receive the Tutti-poles. Tutty means a nosegay, and the poles have a garland at the top. Each is about the size of a broomstick, handsomely decorated with streamers of coloured ribbons and having a nosegay of bright flowers and an orange at the top. It is probable that the custom derives from some pagan phallic ceremony, of which the pole itself is an obvious emblem, while the ' rights ' of the Tutti-men to kisses remind us of the considerably wider opportunities enjoyed by menfolk during certain pagan festivals.

However, in these days everything is done decently and in order. With the Tutti-men goes the Orange Scrambler, whose hat is decorated with cock's feathers, and whose bulging sack will be several times refilled and scrambled during the day. Behind follows an excited crowd of children – *they* know who will get the oranges.

THE ORANGE SCRAMBLER. THE TUTTIMEN AT WORK.

HOCKTIDE AT HUNGERFORD.

To face p. 40.

The Tutti-men make the round of the town, and by the ancient law they are entitled to demand a kiss from every female they meet, or in default a fine of one penny is paid by the reluctant fair. I have followed them closely, but I have never seen any money paid.

Their first call is at the Workhouse, where they distribute tobacco, snuff, and oranges among the old folks, and kisses galore for the pretty nurses. They are then required to call at the house of every commoner in the town. If there are any ladies in the house, each receives a kiss and an orange from the top of the pole ; or in default pays the usual fine. The common-right houses number about a hundred, and it may be that they were only required to kiss the ladies in these houses ; but in practice the custom has developed into demanding a kiss from every lady. I have even seen motor-cars stopped, in order that the pretty driver should pay her due. The Tutti-men are handsome young fellows, and I don't think they get much in the way of fines.

Meanwhile the Court of Feoffees has assembled in the Town Hall and elected the officers for the year, Constable, Portreeve, Bailiff etc.,[1] and has dealt with all the business for the session ; and at 1 p.m. all assemble at the Three Swans Hotel for the annual luncheon. Though the exterior has

---

[1] A word of explanation as to these ancient offices may be of interest. The Bailiff collects the tolls and summons the jurors, and on his retirement becomes Portreeve and collects quit rents, and is in succession for the office of Constable.

been modernised, this is a fifteenth- or sixteenth-century building, and some interesting ancient panelling and frescoes have recently been discovered.

At the luncheon the newly elected Constable is an ex-officio Chairman and the company includes the Feoffees, the commoners, the Tutti-men and most of the notabilities of the town and district – also one nonentity – this writer. The Tutti-poles make a splash of colour on each side of the Chairman, and the April sunshine forms a halo round his head. It is a good omen – men say – when the sun shines on the new Constable.

A solid luncheon having been demolished, the time has come for a bowl of punch, and lighter matters, to wit ' Shoeing the Colt.'

The loyal toasts have been drunk, and on the table are long churchwarden pipes by the bushel for those who favour them, with the more modern cigars or cigarettes for those who do not.

The Chairman rises.

" Gentlemen " (yes, *gentlemen*, NOT ladies and gentlemen ; this is one of the few remaining preserves of the *weaker* sex), " I am informed that there are strangers present, so, according to our rules, *we must shoe the colt*."

A tall stout gentleman appears, wearing a blacksmith's leather apron, and with his coat off and shirtsleeves turned up. He has a hammer in his hand, and by his side is an assistant with a box of farrier's nails. In private life he is a

SHOEING THE COLT.

To face p. 48.

highly respected local magistrate, now he is the blacksmith. "First Colt, please."

The Vicar, by right of his cloth, is the first victim (see photograph). The blacksmith seizes his foot with both hands, passes it between his legs, and holds it fast. He then takes a nail from the box, and pretends to drive it into the boot-heel, giving mighty blows to the sole of the boot meanwhile. This is continued until the victim cries " Punch " and is freed, but pays his score of 10/- to the Chairman, which goes for drinks to the company.

The next victim was an individual who, as a layman, and worse still a writer, was deemed fit for rougher handling. As he totters insecurely on one leg, bystanders jostle him till it is difficult to stand upright. The company – justly indignant – regard this as evidence of viciousness, and a perfect chorus of " Steady, Boys," Whoas, and Clicks resound ! The blacksmith soothes the restive steed by patting his neck (or pulling his ear) and the hammer-blows continue until the victim purchases his freedom at the standard rates. Later a stranger objects, he does not wish to be 'shod.' The Chairman rises, with stern official warning.

" Remember, we have power, by our ancient rules, to fine defaulters one pound." The objection is withdrawn, and the fun continues.

It might be explained that 'strangers' include not only visitors from a distance, but also fresh commoners who may have come into the town during the year, and residents who are not old inhabitants.

Meanwhile the children have been anxiously waiting outside until the Tutti-men have finished their luncheon ; then all the remaining oranges are scrambled and the youngsters go away with bulging pockets.

Readers who wish to see this interesting survival of the past should note the following facts. The date is the Tuesday in the week following Easter Monday. The hornblowing is at 8 a.m., the Tutti-men's parade starts at 9 a.m., and the orange scrambling, after the luncheon, is about 2 to 2.30 p.m. I will close by quoting a few lines by Mr. A. W. Neate setting forth the privileges of the town, of which all the inhabitants are justly proud.

> ' Great Alfred, whom the foe so feared,
>   By him our land was freed,
>   E'en yet good Alfred is revered,
>   Who taught the folk to read.
>   And John O' Gaunt's remembered too
>   By local lucky wights ;
>   They bless his name, the favoured few,
>   Who have their Common Rights.
>   The annual Hocktide Holiday
>   Our institution is,
>   For oranges we romp and play
>   And think it quite " good biz,"
>   But sure the maids appreciate
>   The privilege that's theirs,
>   As they so anxiously await
>   The Tutti-men in pairs.'

The reference to King Alfred is due to the fact that he was born at Wantage, close by, and is supposed to have won one of his greatest victories on the downs not far from the town.

EASTER CUSTOMS

TO the devout Christian the festival of Easter is of supreme importance because it commemorates the Resurrection of the Redeemer, and everywhere the Church observes it with solemnity, and often with gorgeous ceremonial. I know of no place in the whole world where anything can be seen to equal the Santa Semana festivities at Seville. They continue for a full week and attract crowds of tourists from all over the world. The processions are specially fine, the Nazarenos are cloaked and hooded like familiars of the Holy Inquisition, and follow gorgeous Baroque Pasos through the Sierpes and across the Plaza to the splendid Cathedral. Each Brotherhood has its distinctive attire, but all are in mourning or semi-mourning colours, such as black, red, purple, cream, dark blue, or combinations of two of these. The cross-bearers are usually barefooted, and all trudge wearily through the streets the whole night.

Similar scenes can be seen at Malaga and Madrid; and there are also impressive ceremonies at Rome and Jerusalem.

Though the Church has decreed that Easter is the greatest of all her festivals, the very name is pagan; and so doubtless is the origin of this festival of Spring. The word Easter comes from Eostre or Ostara, the Anglo-Saxon goddess of Spring, and it is probable that when the heathen Saxons became Christian, their festival of the

Spring goddess became the Christian Easter. There is reason to think that this is what happened elsewhere, so that Easter is a Christian adaptation of former pagan Spring festivals.

The Founder of Christianity Himself observed the Jewish Passover, as did the Jewish Christians; and we find an interesting reference to this in the words used for Easter by European Christians. Thus in French Easter is Pâques, Italian Pasqua, Spanish Pascua, and Dutch Paach, all of which come from Latin Pascha and Greek πασχα (pascha) from the original Hebrew Pesach, a verb meaning 'He passed over.' Thus to the early Church Easter gradually replaced the Passover feast, and, like it, was a lunar date; Easter Day being the first Sunday after the full moon following the vernal equinox. Owing to variations of latitude and the inaccuracy of early calendars, there have been many disputes as to the correct date of Easter, one of which all but rent the Church in twain. This was the quartodeciman controversy between the Eastern and Western Church, and to this day the former observes a different date from the Western world. In modern times we have seen the growth of a movement for a fixed Easter, and it is probable that the practical convenience of the arrangement will ultimately gain the day; especially as the growth of motoring and the cult of outdoor life are rapidly transforming its character. Every year to the masses of the people it becomes more a spring holiday and less a solemn religious feast.

In this country we have few survivals of the gorgeous ceremonial of the mediæval Easter. Perhaps the most interesting and picturesque of all is the Maundy distribution at Westminster Abbey. This takes place on Maundy Thursday, *i.e.* the day before Good Friday, and it is supposed to commemorate Christ's washing of His disciples' feet. In former times the Sovereigns of England did actually wash the feet of a baker's dozen of beggars, which, however, had undoubtedly received a preliminary cleansing process from some minor official.

The last monarch to perform the full ceremony was James II, no subsequent sovereign having continued the Feet-washing ritual, which is of Catholic origin. Until the result of the Great War deprived Austria of her throne, it was the custom for the Emperor to perform the ritual, and it is still observed by King Alfonso at Madrid.

In England, on the other hand, since the time of James, our sovereigns have contented themselves with the distribution of Maundy Money, but vestiges of the Foot-washing can still be seen in the white towels worn both by the officials and by the recipients of the bounty. It is an impressive affair, and I was fortunate enough to gain admission to the Jericho Chamber at the Abbey where the preliminary arrangements are made. The Dean puts off his golden vestments and wears a simple white sheet or towel, as does the Lord High Almoner. We were in the 63rd year of King George, and so, according to the ancient rule, there had to be sixty-three recipients of the royal bounty, and a like number of pennies given away. Actually, for reasons of space, only about half were present, but among them I noticed four children (two girls and two boys) wearing the white towels and carrying posies of flowers ; and certainly none of them needed the ceremonial washing which is now omitted.

The sixty-three pennies, made up from a currency specially minted for the occasion, in denominations of 4*d*., 3*d*., 2*d*., and 1*d*. and all *in silver*, are contained in small red and white leather purses to which long strings are attached. These were the usual purses of mediæval times, and illustrate the real meaning of the proverb 'Tightening purse-strings,' which to-day is utterly pointless. They are placed on a great golden dish, with the strings hanging round in a fringe, and woe betide the patient official who so arranges them if the strings become entangled when the Great Personage distributes them !

Though 5/3 may seem a paltry gift for a King, actually these tiny pieces of money are greatly prized by collectors and offers of £1 or more for the set are quite common ; and Treasury notes for a more substantial amount are often added. Very interesting are the two splendid Yeomen of the Guard who are in charge of the dish. These are our old friends the Beefeaters of the Tower, but are wearing to-day their magnificent full uniform of scarlet and gold, with Tudor rosettes on shoes and hats. In their belt is a large, awkward, iron spike or swivel, placed there to support the heavy arquebus of three hundred years ago.

Of the two Yeomen of the Guard, one carries the golden dish, while the other is a sergeant-major with staff of office, who accompanies him. Originally this was to prevent his bolting with the money, but I need hardly say that this is now a totally unnecessary precaution. All the Yeomen who take part in the ceremony are picked men, the two fine old veterans in my photograph having seen much war service, and it is a matter of great pride to them to be selected for the duty. The actual distribution of the money takes place in Westminster Abbey, usually by the King or his deputy, the latter being generally a member of the Royal Family.

The service takes place at 12 noon on Maundy Thursday, and the public are freely admitted to the Abbey to see this interesting survival of mediævalism.

There are several interesting old customs still observed on Easter Monday. The most remarkable of these is the celebrated Biddenden Dole, which is given away at the White House, Biddenden, Kent, at 10 a.m. The Dole consists of a distribution of bread, cheese and ' Biddenden Cakes,' to those who come up for their share. The village is about $13\frac{1}{2}$ miles from Maidstone,

A BIDDENDEN CAKE.

THE KING'S MAUNDY CEREMONY.
Note gold dish of purses.

To face p. 56.

and less than fifty from London, so there is usually a little crowd of visitors from a distance.

The bread and cheese are, of course, reserved for the poor of the district, but the highly interesting cakes are freely given to the visitors who come to see the ceremony, and form a fascinating souvenir. They are not cakes, but biscuits about four inches long, and are stamped with a representation of the celebrated Twins, the givers of the Dole. The legend states that about the year 1100 twin girls were born in the village. Their names were Eliza and Mary Chaulkhurst, and they were joined at hips and shoulders, like the more recent Siamese Twins. They were quite healthy young women until they reached the age of 34, when one of them became seriously ill, and soon after died. Her surviving sister refused to be separated from the corpse, using the pathetic words, " As we came together, we will go together," and some six hours later, in spite of all care and attention, she also was taken ill and died. Their place of interment is still pointed out in the Parish Church. They bequeathed twenty acres of land to the churchwardens of the parish, the rent of which was to be expended in bread and cheese for the poor of the parish. These lands are still pointed out and are known as the " Bread and Cheese " lands ; while the revenue from them furnishes the yearly " Dole." In 1636 a greedy Rector attempted to add the lands to his Glebe, but happily lost his action.

There seems no reason to doubt the truth of the main features of the legend, but the year 1100 is certainly incorrect, and the date was probably about 1560.

The 'Cakes' have a very quaint representation of the sisters stamped on them. Their names are printed at the top, on their aprons is graven their age (34) and the reputed year of their death (1100). On the village green is a handsome Village Sign, taking as its emblem the celebrated Twins who have made the village famous.

There is another Easter Monday Dole at Ellington (Hunts), which I have described fully in the chapter on the Tichborne Dole; and it should not be forgotten that though these may seem merely picturesque survivals to-day, they must have been a valuable aid to the poor in the bad old days, when the poor were really poor.

The old town of Preston has a long corporate history and is proud of its adherence to old customs. One of these is still carried on and can be seen on Easter Monday. This is the old Egg-rolling game. Large numbers of people assemble to watch it or take part. The eggs are brightly coloured, in many different hues, and are rolled down a hillside. In mediæval times this was said to be symbolical of the resurrection, as rolling the egg represented rolling away the stone from the tomb of Our Lord. There can be little doubt, however, that the rite is older than Christianity and was originally connected with the Spring Festival. This was found in nearly all lands, and was connected with the fertilisation idea; the egg is an obvious symbol of the life force.

The children in parts of Yorkshire also practise egg-rolling at Easter, and there is a lane near Barton-on-Humber where they roll the eggs until the shells are broken. The common Easter Eggs, which are sold everywhere at Easter, of course are connected with the same cult, and represent the idea of the renewal of life and the rebirth of nature.

There is another old custom, which now has nearly or quite died out. This is Easter ' Lifting ' or ' Heaving,' which is said to have been designed to commemorate Christ's Resurrection ; *i.e.* the raising of the body from the grave. It was chiefly observed in the northern parts of the country, particularly Shropshire, Cheshire and Lancashire on the Monday and Tuesday after Easter. On the first day the men ' heave ' the women, and on the second the latter perform the same office for the men.

A party of men go round with a chair, force every female to be seated therein, and raise her three times with loud cheers. For this they claim a chaste salute, or a fine of one shilling in lieu thereof. The next day half-a-dozen women go round and perform the same service for the men. This custom led to so much rowdy horseplay that it has been largely stopped through the influence of the magistrates. There is a record of one King (Edward Longshanks) having suffered the indignity, being lifted bodily from his bed on Easter Monday morning by a party of lively maids of honour.

April 1st

ALL FOOLS' DAY

> 'The first of April, some do say,
> Is set apart for All Fools' Day.
> But why the people call it so
> Nor I nor they themselves do know.'
> (From *Poor Robin's Almanack*, 1760.)

THE custom of playing practical jokes on the first of April can be traced over most of Europe, with the single exception of Spain. The commonest form of the joke is to send the simple-minded on some fruitless errand, children naturally being the easiest victims. In France they are sent to get a dozen cocks' eggs, or a stick with only one end. In Brittany they go for a litre of sweet vinegar. In England they are sent for a pound of elbow-grease, a leather hammer, or a quart of pigeons' milk. In Scotland, during the 1st and 2nd of April, it is the custom to send a simple-minded person to another with a note, who on opening it informs the bearer that it ought to have been delivered to another person. This joke is kept up as long as possible, and when the infuriated bearer of the letter opens it, he finds it only contains these words

> "The first and second of Aprile,
> Hound the gowk another mile."

Gowk is the Scottish for cuckoo, which is the

term applied to April Fool, while in France he is the ' Poisson d'Avril ' or April fish.

Though the joke is of exceedingly ancient origin, there is no certain explanation, but it is possible that it was derived from the Jews. A writer in the *Public Advertiser* of April 13th, 1769, suggests that the idea of sending a person on a fruitless errand is derived from the dove sent out by Noah from the Ark, which was obliged to return, ' finding no rest for the sole of her foot.' He says that this happened on the day of the Hebrew month which answers to our 1st of April. Another suggestion is that there was an ancient Jewish custom of sending an unpopular person from one house to another ; and that is the meaning of the action of the Rulers of the Jews in sending Christ on to Herod, Annas, Caiaphas and Pilate.

It is more likely that the custom relates to some primitive usage at the vernal equinox, especially as on that date the sun emerges from the zodiacal sign of the fishes. This no doubt explains the French term ' Poisson d'Avril.'

April 24th

VIGIL OF ST. MARK: MIDNIGHT WATCH IN CHURCH PORCH

THE evening of April 24th is St. Mark's Eve, and it was formerly believed in rural districts that any person who has the courage to do so can wait in the church porch and see the spirits of all the people who will die during the next twelve months. There are, however, two serious risks attached to this interesting investigation. It is, tradition avers, essential to keep wide awake, for if the watcher falls asleep he will never wake up again; and, moreover, if the practice is once commenced there is no getting out of it. It must be done every year to the end of life.

A writer in the *Evening Standard* of April 24th, 1929, says that an old countryman in Yorkshire spoke thus about it:

'I remember very well hearing folks talking about it years ago. I never watched myself, but there was one old fellow in Wensleydale who used to sit in the church porch to watch the dead march into the church every St. Mark's Eve. He had to do it, he couldn't help himself. He'd done it once, and he had to go on with it. They said he used to see the spirits of all the people who were going to die in the next year. Of course, it happened at last that he saw himself in the procession. He must have done. At any rate, he was dead and buried before another twelve months were out.'

The same writer reported that Mr. Elliott O'Donnell had thus spoken of the old custom :

'I have heard of sceptics putting it to the test, and of " singing to a different tune " when the phantasms of those they knew quite well suddenly shot up from the ground, and gliding past them vanished at the threshold of the church.'

Brand in his *Popular Antiquities* says that the custom of watching in the church porch was general in Yorkshire, and that it was usual to wait from 11 p.m. till 1 a.m. (This was written in 1841.)

He also says that many sick persons have actually died from their fears, set up by the statement that their ghost had been seen at the church porch on the previous St. Mark's Eve and by their consequent belief that they must die within a year.

It is not with death alone that the Vigil of St. Mark is concerned. A century or so ago it was the custom of love-lorn girls to seek to discover their future husbands at midnight. *Poor Robin's Almanack* tells us the method :

> 'On St. Mark's Eve at twelve o'clock
> The fair maid will watch her smock,
> To find her husband in the dark,
> By praying unto good St. Mark.'

MAY

MAY DAY CEREMONIES

THE FURRY DANCE, HELSTON, CORNWALL

WELL-DRESSING CEREMONIES IN DERBYSHIRE

WELL-DRESSING AND HOLY WELLS

THE MORRIS DANCERS

THE FESTIVAL OF THE DUNMOW FLITCH

WHIT-MONDAY AT SOUTH HARTING, SUSSEX

THE KNUTSFORD FESTIVAL

MAY DAY CEREMONIES

'You must wake and call me early, call me early, Mother dear;
To-morrow 'ill be the happiest time of all the glad New Year;
Of all the glad New Year, Mother, the maddest merriest day;
For I'm to be Queen o' the May, Mother,
I'm to be Queen o' the May.'

I MUST apologise for heading this chapter with such a trite quotation, but I know of no other which so well represents the spirit of the Merry Month, and the one which is the richest of all the year to the student of folklore and the lover of the beautiful or picturesque.

In my boyhood I well remember the children coming round with garlands to every house, singing:

'The first of May is Garland Day,'

and though, unfortunately, modern rush and compulsory school attendance have almost killed this ancient custom, we are still entitled to regard this as the garland month. And when we repeat the words 'The first of May is Garland Day' we must remember that before the alteration in the calendar, the first of May came almost a fortnight (actually eleven days) later and so flowers and blossoms were more plentiful. This explains why it is that we hear of hawthorn blossoms being carried in May Day garlands, when they are never obtainable to-day. In mediæval times the young men and maidens were wont to wander into the woods during the greater part of the night, returning at daybreak with garlands of flowers or branches of trees.

Shakespeare alludes to this in Midsummer Night's Dream :

> '. . . If thou lov'st me then,
> Steal forth thy father's house to-morrow night,
> And in the wood, a league without the town,
> Where I did meet thee once with Helena,
> To do observance to a morn of May,
> There I will stay for thee.'

There can be little doubt that this ancient custom originally derived from some pre-Christian usage, when in Spring certain laxities were permitted. Some vestige of these certainly continued in mediæval times. Stubbes, in his *Anatomy of Abuses*, published about 1583, says :

'Against Maie every parish, town and village assemble themselves together, bothe men, women and children, olde and young, even all indifferently : and either goyng all together, or deuidyng themselves into companies, they goe some to the woodes and groves, some to the hilles and mountaines, some to one place some to another ; where they spende all the night in pastimes, and in the mornyng they returne, bringing with them birch bowes, and braunches of trees to deck their assemblies withall.

'And I have heard it credibly reported, and

that *viva voce*, by men of great gravitie, credite and reputation, that of fourtie, three score, or a hundred maides goyng into the woodes ouer night, there have scarcely the thirde part of them returned home again undefiled.'

There is an ancient custom still regularly observed at Oxford and Southampton, in the form of an early morning religious service at the top of a tower. At Oxford it is held at 6 a.m. on the top of Magdalen Tower, when a brief Latin service is held, and a surpliced choir sing several hymns. I have attended it and found it very impressive; while a crowd of at least a thousand people waited in the street and on Magdalen Bridge to see and hear all they could. To-day it is performed with reverence and decorum, but old residents of Oxford remember the days when it was considered an excellent opportunity for a ' rag ' by the 'Varsity lads, who from the security of the summit of the towers, or from the battlements of the college, were able to throw down rotten eggs and bags of flour on the crowd in the street below. A similar service is still performed on the top of the ancient Bargate at Southampton at the same hour.

Both of these date from mediæval times, and the former is said to have originated in a Requiem Mass for the soul of King Henry VII, which after the Reformation was altered to the musical service, for the maintenance of which the Rectory of Slimbridge (Glos.) pays annually £10.

In the case of Magdalen Tower, the vantage-point chosen is the most eastern in the city, though of course the Tower itself only dates from the fifteenth century; but the original usage may be a memory of something very much earlier. Sun-worship was always from a lofty situation.[1]

There can be no doubt that the Beltane customs in Scotland on the first of May, and the Bonfires on Midsummer Eve, do actually date from the pagan worship of Baal and of the human sacrifices which characterised that dreadful cult. The Beltane usage on the first of May, which was observed up to the eighteenth century in Scotland, consisted in parties drawing lots from a hat, the person who took the black piece being he who was to be sacrificed to Baal; though he was only then required to leap three times through the flames of a bonfire. A very similar usage on the first of May was formerly observed in Munster and Connaught, when the herdsmen drove their cattle through fires, to preserve them from disorders during the coming year.

Fascinating though these speculations are, the main purpose of this book is to deal with old customs which can still be seen. The first of May

---

[1] In the past the world was supposed to be flat, and so naturally any hill or mountain was believed to be nearer to the Sun-god; and in very flat country lofty towers were erected. E.g. Babel. Hill-tops were therefore considered to be the most sacred parts of the earth, for there the gods descended in clouds and had their interviews with men. We see this idea in the story of Moses on Mount Sinai, and in the Greek mythology in which the gods resided on Mount Olympus; and all through the Old Testament we have constant reference to worship in 'high places.'

This indicates that a religious service held on top of a tower may have some connection with Sun-worship, even though long forgotten.

is the day for May-poles and May Queens. I am delighted to say that this most picturesque old usage is still carried on in a large number of towns and villages throughout the country, and I have personally visited many of them. Strangely enough, London itself (though sadly lacking in wild flowers) is very zealous in keeping up the Crowning of the May Queen; and in addition has a most remarkable festival at which many May Queens attend, the prettiest being crowned as 'May Queen of London.' Most of the London Elementary Schools have their own festival on May Day, and a May Queen is selected, crowned and attended by many pretty children carrying garlands of flowers.

On the first Saturday after May 1st (or the same afternoon, when May 1st falls on a Saturday), the great May Queen of London Festival is held on Hayes Common, near Bromley, Kent. It is a truly delightful festival of youth and beauty. It usually commences about 2 p.m. with a procession of May Queens, attended by Maids of Honour carrying garlands, and many little tots with gaily decorated prams containing May Dolls. They proceed to the Common, where the prettiest of the May Queens is crowned as 'May Queen of London.' The ceremony is performed by 'The Prince of Merrie England,' who is a pretty girl, dressed in tights like the principal boy in the pantomime; and is usually the May Queen of London of last year. It is an extremely attractive sight, the gay costumes of the little

THE MAY QUEEN OF LONDON AND THE PRINCE OF MERRIE ENGLAND.

THE MAY QUEEN AND HER COURT.

MAY DAY CEREMONIES.

To face p. 70.

girls, the ribbons, and the flowers forming a lovely picture. The May Queens are usually very pretty girls of twelve to fourteen years of age ; and even if the May Queen of London only wears a tinsel crown, she has that which very few real queens possess – radiant youth and beauty. The May Queen, having been duly crowned, receives the homage of all the lesser queens, who sweep up to her, make a deep curtsey, and retire again. A large choir gives musical features at intervals, and there is dancing round the Maypole, and ' all the fun of the Fair.'

The same custom is also kept up with great ceremonial at Knutsford, Cheshire. At Minehead on May Day the fishermen make a cardboard ship, ten feet long, the sails trimmed with ribbons. This is carried by a man, and to the stern of the ship is fastened a cow's tail, and people who will not put their money in the collection box are threatened with a beating from this. Local tradition states that the custom dates from a shipwreck in the eighteenth century, when a vessel was sunk off the coast with all hands ; the only survivor being a cow which swam ashore.

At Barwick in Elmet, near Leeds, there is a very remarkable festival held once in three years (the next will be in 1931), when the pole is taken down and re-painted and the four garlands renewed. These are remarkable affairs of rosettes, ribbons and artificial flowers made by the Girl Guides, and built up on a wire foundation eighteen inches high. The pole is taken down on the Monday in Easter week, and set up again on Tuesday in Whitsun week, by the traditional method of ropes, pitchforks and ladders. Three men are elected as Pole Men and are responsible for all arrangements. Some of these poles have been as much as 95 feet high, but the present one (1929) is only 80 feet. There are the usual festivities, and the villagers are very proud of their ancient festival.

The remarkable garlands at Barwick are very similar to those still in use at Wadworth near Doncaster, which are plaited from coloured ribbons by the old ladies of the village. One of the most remarkable May-poles still extant is at Welford, near Stratford-on-Avon. It is exceptionally lofty, and is painted in spiral red stripes like a barber's pole. It is interesting to mention that the sign of the Barber's Pole was intended to represent a bandaged limb, for the barber—or leech—was the original surgeon.

Some other May-poles can be discovered at Wellow, Redmore, Donnington, Preston Brockenhurst, Hemwell, and Wanstead. There is another on the village green at Temple Sowerby (Westmorland), which the Lord of the Manor is required to replace when it decays. There is another at Shillingstone (Dorset), but the date of festival is June 9th, and the reason is interesting

Originally the date was May 1st, like all the other villages, but in a burst of loyalty at the

Restoration of Charles II and the recommencement of the innocent rejoicings which had been banned by the Puritans, the folk of Shillingstone decreed that henceforth the festival should be held on May 29th. This was not only the birthday of the 'Merry Monarch,' but also the day upon which he entered London, having landed at Dover five days earlier. Presently, however, the change in the calendar took place, and the date jumped forward eleven days. Hence June 9th.

As to the origin of the May-pole, there can I think, be no doubt at all that it is a pagan phallic symbol and that the Pole is the lingam. The Puritans regarded it as an idol, and prohibited its use by an Act of 1644, but they were, of course, ridiculed by the populace. An old song published in 1634 says :

> 'Some fiery zealous brother, full of spleene,
> That all the world in his deep wisdom scornes,
> Could not endure the May-pole should be seen
> To wear a coxe-combe higher than his hornes :
> He tooke it for an Idoll, and the feaste
> For sacrifice unto that painted beast ;
> Or for the wooden Trojan asse of sinne,
> By which the wicked merry Greeks came in.'

The laxity permitted to the maidens of the village on the night previous to May Day, and the connection of the Pole itself with the phallus or lingam, strongly points to pre-Christian festivities relating to the fertilisation idea ; and it is probable that the mediæval May Day ceremonies are partly based on some such forgotten pagan usage. Of course, it was only a faint memory, since all the crudities and grossness of the earlier rites would have been prohibited by the Church and forgotten by the people.

The custom of the Jack in the Green, which was till recently observed in connection with May Day ceremonies, is of interest. The sweeps turned out in great force, wearing grotesque dresses and carrying the inevitable collection boxes. In their midst was the 'Jack in the Green,' a man in a wicker framework, completely enclosed in leaves and shaped like a pyramid. It is only a suggestion, but this may have some connection (as a far-off memory) with the brutal custom of the Druids, who constructed a wicker framework in the shape of a man, crammed it with living persons and set it on fire. In this connection it should be remembered that the great festival of the Druids at Stonehenge took place in summer, and was no doubt connected with human sacrifices (see 'Midsummer Morning at Stonehenge'). This may seem far-fetched, but my study of folklore has convinced me that many of these old customs really go back to origins beyond the dawn of history; and the careful searcher will find embedded in them – like fossils in the rock – vestiges of primitive usages which provide precious clues to the customs, habits and beliefs of far-distant ancestors whose very names have been lost for ever. And as the scientist can construct the mighty plesiosaur or the gigantic cetiosaur from a few bones found in the oolites, and a great forest from a few fronds discovered in a lump of coal, so the student of mankind can conjure up a picture of bloody human sacrifices, or gross and licentious phallic orgies, from some stray word or symbol which remains to us to-day in some innocent country dance, or ancient folk-festival.

And if this seems too wild a guess and too bold a claim, consider this. Recently I was speeding in my car through a typical stretch of our beautiful homeland. I passed through fields of ripe corn, and rich pastureland dotted with contented cows; I saw sweet villages, peaceful churches and splendid mansions; and yet – the record of the rocks tells us that once that very stretch of countryside was a tropical swamp and that through its tepid waters waddled and splashed gigantic reptiles bigger than a motor 'bus, creatures that could pull down and devour a great tree. Instead of the spreading oaks and shady elms, weird and strange tree-ferns lifted their stiff fronds towards the steamy sky, and huge flying-dragons, with beaks full of teeth, flew and screeched, and fell upon and devoured lesser creatures.

May 8th

THE FURRY DANCE, HELSTON, CORNWALL

OWING to its isolated geographical position and special racial characteristics, the Duchy of Cornwall is particularly rich in picturesque folk survivals. The most interesting and significant of these is the celebrated Furry Dance, which is still kept up with the utmost vigour and enthusiasm. The people get up early and go to the fields to collect garlands of flowers and green branches, and deck themselves therewith. Then they dance in the streets in couples and go right through the houses, in at the front door and out at the back. This is the speciality of Helston, and the inhabitants leave their doors open obligingly for the dancers. As with the Morris Dancers and Mummers, a special tune is played, which has been handed down by oral – or rather *aural* – tradition from time immemorial; and the same is found in both Wales and Brittany, thus clearly pointing to a very early common origin. They also sing the Furry Song, mentioning both Robin Hood and the Spaniards, which has led some to think it dates from the Elizabethan period. It goes with a swing like this:

Robin Hood and Little John,
They both are gone to Fair O,
And we will to the merry Greenwood
To see what they do there O,

And for to chase O,
To chase the buck and doe.

Whereas those Spaniards that make so great a boast O,
They shall eat the grey goose feathers, and we shall eat the roast O
In every land O.   The land that e'er we go.

Chorus.

With Hal-an-tow ! Rumbelow !
For we are up as soon as any day O,
And for to fetch the summer home,
The summer and the May O.
The summer is a come O,
And winter is a gone O.'

In 1907 Sir William Treloar, who was a freeman of Helston, led off the dance down the streets with a local young lady, and ended up with a waltz in the Assembly Rooms. It should be mentioned that the fun begins at a very early hour, and nobody in Helston gets much sleep after 4 or 5 a.m.

The custom is supposed to be derived from the Roman goddess Flora, whose festival was celebrated by revels known as the Floralia. This lady did not commence life as a deity, but as a courtesan, and when she died bequeathed her whole fortune to the people of Rome so that they might annually celebrate her memory in May. The grateful Romans elevated her to godhead, and certainly she was quite as worthy as some of their Emperors who shared the Olympian Heights with her. Her festival was, of course, simply a saturnalia and would doubtless have been characterised by the removal of sexual inhibitions; such festivals were common among the Romans, Greeks, Phœnicians and Hellenistic peoples of the Mediterranean basin in classical times, and there is reason to believe that all primitive peoples at some period have observed this practice. It is well known that in classical times a woman was compelled, once in her life, to give herself to a stranger, and the saturnalia was an occasion to pay this debt of (dis) honour. I am of opinion that the Furry Dance at Helston and the Hocktide Festival at Hungerford are vestiges of such a custom here. Certainly the ancient obligation to open the doors of private houses is most suggestive. Two thousand years ago sex was neither hushed up nor suppressed. We have abundant proof of this in the Latin and Greek poets, and a walk through the streets of Pompeii, Timgad or any similar city even to-day would probably be fatal to Mrs. Grundy. I have seen the crude but unmistakable emblem graven on the pavement in Pompeii, and on house-fronts at Timgad, and it appears in frescoes and mosaics as a popular lucky sign, and was even used for door-knockers and brass table ornaments. We find it both in sculpture on the temples of India and among the mysterious ruins of Central America, the lingam magic talisman of the life-force.

Some writers derive the name Helston from Hellstone, and appropriate legends are readily coined to fit. It seems quite possible to me that the word may have been originally Hele-stone (or sun-stone), and that an upright pillar for worship stood here like the Friar's Heel at Stonehenge. Nothing can now be proved, as no such stone exists.[1] Cornwall, however, is exceedingly rich in megalithic monuments, and single pillars are not uncommon. There are many suggestions as to the origin of the word Furry. Mr. J. Harris Stone derives it from Celtic ' feur,' a fair holiday ; an old parochial history published in 1838 says it is simply ' foray,' and Mr. Lewis Hind says it is a corruption of Flora Day.

[1] The most interesting of the Hellstone legends is to the effect that a fight took place between St. Michael and the Devil in mid-air over Helston. The Evil One was getting the worst of it, and dropped into the streets of the town a block of granite which he had brought from Hell. Dr. George C. Williamson (from whose book I quote this legend) says that the stone was broken up in 1783.

WELL-DRESSING CEREMONIES IN DERBYSHIRE

ONE of the most beautiful of all the old customs which are still annually observed in England is Well-Dressing. The best place at which to observe it is in the beautiful Derbyshire village of Tissington, on Holy Thursday (or Ascension Day). The village lies about half a mile off the main Derby to Buxton road, and is about 4 miles from Ashbourne, 16 from Buxton, and 17 from Derby. Motorists proceeding northwards from Derby will look out for a gate on the right about $3\frac{1}{2}$ miles after leaving Ashbourne; though closed by this gate, the road is public and leads to the village.

There are five wells in the parish. The first we come to is called Yew Tree Well (formerly Goodwin's Well), and in close proximity are the Town Well, the Coffin Well (so called from its shape), Hand's Well (so named from the family who lived for many years at the adjacent farm), and the Hall Well, opposite the gates of the Hall, which is the largest and finest. All supply delicious clear fresh water, at a temperature of 47 degrees Fahrenheit, all the year round. In the Black Death, 1348–9, when more than half the inhabitants of the county of Derbyshire died, Tissington alone escaped, and this fortunate circumstance was ascribed by the grateful inhabitants to the mercy of Almighty God and the particular purity of the water.

Hence the founding of a festival, which was

first held on Ascension Day 1350, when the plague generally died out in the county; and which took the form of a service of thanks to God, and some kind of dedication of the wells, which were believed to have been His instruments in saving the people. It is very likely that the remarkable purity of the water greatly helped the people during that trying time, but I cannot ignore the fact that the village lies off the main road, in a sheltered yet lofty position on the hill side. Here they had pure water and air, and little or no risk of infection being introduced by travellers, owing to the remoteness of the village and the extreme badness of the roads.

WELL-DRESSING AND HOLY WELLS

WHILE I think it is very likely that an *ancient custom was revived* by the Well-Dressing festival at Tissington in 1350, it is highly improbable that it originated then. Well-Dressing ceremonies have been, or are still, carried on in other places, where no legend of the Black Death exists to account for it; e.g. Baslow, Buxton, Wirksworth, and Belper, all in Derbyshire, and in some places in Staffordshire. There can be no doubt that this is a survival of the simple nature worship of the earlier inhabitants of our country, and as the matter is of great interest, I propose to devote some space to it later.

Now I must return to our Well-Dressing. Briefly, the ceremony consists in decorating the wells and afterwards holding a brief service before each of them. The decorations are wonderful, being the handiwork of amazingly skilled craftsmen working with the simplest materials. Large wooden frames are spread with clay, which has salt water mixed with it, to keep it moist for several days. The surface of the clay is smoothed perfectly flat, and upon this unpromising canvas these rural craftsmen construct wonderful pictures in the most vivid natural colours, using the simplest materials. Grains of rice for white, the petals of bluebells for blue, red and pink daisies; wonderful shades of green are obtained from young larch buds, and

YEW-TREE WELL, TISSINGTON.   HAND'S WELL, TISSINGTON.
HOLY WELLS.   Note the picture in flower petals.

To face p. 82.

darker shades from mosses, greys from lichens and sometimes brown and red berries are used. Really ambitious pictures are constructed of glowing colours that put to shame the finest tints of the artist's brush, and thanks to the moistened clay they keep fresh for three or four days. This wonderful craft has been handed down from generation to generation, and the results are simply wonderful. Each picture is contained in an ornamental border, with a Bible text at the top.

On the occasion of my own visit I observed the following designs. At Hand's Well, a finely drawn picture of an angel with a floral background, and the words 'Great is Thy power'; at Yew Tree Well a fine rendering of a cherry tree in bloom and the words 'Praise ye the Lord'; at Coffin Well a very fine design and the words 'Thy faith shall be justified'; and at Town Well perhaps the best of all. It was a really magnificent landscape, having a mountain stream in the foreground with a stag drinking and in the middle distance a splendid mountain peak, whose craggy sides and tree-clothed slopes were wonderfully executed. Except that the colours were fresher, brighter, and more vivid than any pigment, it would certainly be taken at a few paces distance for a water-colour painting.

The religious service is simple, brief and impressive. A procession consisting of the choir and clergyman, followed by a number of people, visits each well in turn, where an appropriate collect and gospel is read, and such favourite hymns as 'Rock of Ages' are sung. The effect is most moving, and the festival is usually attended by a large crowd.

The subject of the 'Holy Well' is a very interesting one. We find Holy Wells and Magic Springs everywhere. I have myself visited four out of the five continents, and in all of them I have found traces of this once universal form of nature worship. It is not surprising that water was regarded as miraculous or supernatural in character, when we consider its amazing effects in arid countries. After travelling for miles through absolutely barren wastes in the Sahara Desert, I have suddenly come upon some oasis whose fertility is in dramatic contrast to the desolation around. A dense mass of lofty palms surround some tiny spring, dry perhaps for part of the year, but having enough water stored to keep the oasis alive till rain comes again. And in the rich soil below the palms, protected by their great fronds from the burning sun, very rich crops of fruit or vegetables are grown in addition to the produce of the palms. Sometimes we find a spring so small that it only serves to nourish a couple of scraggy palms, but even then this is life surrounded by death. And in the brief rainy season, when the hard, sun-baked clay of the Hauts Plateaux is moistened by a passing shower, we get immediately a wonderful growth of tiny flowers, whose almost invisible seeds will sometimes lie for years without germinating, but spring immediately into life when the moisture arrives.

Small wonder that in the childhood of mankind, divine power was believed to be possessed by water, especially living water (i.e. water from a spring), and as it could give life and fertility to vegetation, why not to men? That is why wells were believed to have healing properties, or to be the abode of some nymph or spirit which had the power of granting wishes or answering prayers.

It is a remarkable and most significant fact that we nearly always come across a sacred tree in close proximity with a holy well, and here we are at the heart of that primitive nature worship which may have been the mother of all religions, and traces of which are undoubtedly found in all. Thus in Genesis ii. 9, 10, we read: 'The Tree of Life also in the midst of the garden. . . . And a river went out of Eden to water the garden.' The Sacred Tree and the Fountain of life, what were they? I have said elsewhere that the primitive man, knowing from observation that the lifeforce was passed on through the union of male and female, had reasoned that the whole of creation was formed in this way. The mighty orb of day, blazing in all its strength from an Eastern sky, was obviously the Male, or all-father; and the Earth the female or all-mother; though frequently the moon was coupled or associated with the earth as the Sun's Wife. And the obvious symbol of the Sun-god was the lingam or phallus, the male generative organ. This was frequently represented by a tall straight tree, especially the date palm, but sometimes by a pillar, a cone, or an upright serpent. Equally obvious as a female symbol was a perforated stone, or any cleft or hole in the earth from which water flowed; so that the well was the natural emblem of the earth goddess from whence came the life-giving water.

I have made a close study of the Holy Wells of Britain, and have visited scores of them. They roughly group into two classes—Wishing Wells and Healing Wells, and at the former it is usual to throw a pin or similar object into the water before making the wish – a relic of pagan offerings to the spirit or goddess of the well.

One of the most interesting Holy Wells in all Britain is the Chalice Well or Blood Spring at Glastonbury. The Christian legends are most romantic. Hither, not long after the death of Christ, came Joseph of Arimathea, bearing with him the Holy Grail, or Cup of Blessing used by our Lord at the Last Supper. At Wearyall (now Wirrall) Hill the tired saint leaned on his staff to pray, thrusting the point into the ground, and when his worship was done, lo, the staff had taken root and grown into a thorn tree! This tree was hacked to pieces by a Puritan soldier, who was said to have hewn off his own leg in the process, but cuttings from it can still be seen in various parts of the town, and all blossom at Christmas. The finest of these Holy Thorns is in

THE BLOOD SPRING OR CHALICE WELL AT GLASTONBURY.

SERVICE AT HALL WELL, TISSINGTON.
HOLY WELLS.

To face p. 86.

the Abbey grounds, and there is another beside the Holy Well. But I must return to St. Joseph, who is said to have buried the cup near Glastonbury Tor, and immediately a spring gushed forth and was found to be tinged with the holy blood of the Redeemer, and has ever since shown a reddish colour. Actually a kind of red fungus grows in the well, and tinges the surface of the water with a ruddy hue; therefore this well was regarded as of the utmost sanctity until the Reformation. It was visited by the Saxon and Celtic saints, Patrick, Brighid, Aldhelm and Dunstan, and also many kings. In 1919 a new lid or well-cover was dedicated by the Ven. Archdeacon Farrar, as a thankoffering for peace.

Probably Glastonbury was a holy place long before Christianity. As Avalon, the Isle of the Blest, it figured as the Celtic Heaven. Its sheltered position, cut off by wide marshes and intricate forests from the rest of England, made of it a safe retreat in times of warfare, while the Tor, rising so dramatically from the lowlands, was an obvious venue for Sun-worshippers, since its conical shape suggested the phallic symbol. And here too was a Holy Well and a miraculous tree! Is it not possible that they were venerated for centuries before the legend of St. Joseph began?

The masonry of the well is said by experts to be pre-Roman, and it has been found by measurements made on Midsummer day to be oriented eastwards. This is the same arrangement as at Stonehenge (see 'Sunrise at Stonehenge').

It is impossible to refer to a tithe of the Holy and Magic Wells of England. About twenty-six are said to have sprung up from the spot where some saint was martyred, buried or rested – the commonest form of the tradition being that the spring gushed forth from the spot where the severed head of the martyr fell. Mr. Hope, in his work on Holy Wells, quotes three which gushed forth when the saint or other notability struck the ground with his staff; e.g. Stoke (St. Milburga), Cerne (St. Augustine), and Carshalton (Queen Anne Boleyn); I would venture to add to these Sir John Shorne's well at Marston, Bucks (which, alas, now is rendered so commonplace by an iron pump); and the famous Well of St. Keyne, to which I shall refer later.

Wells to which gifts were brought are very common. I have counted a dozen in Cornwall alone. Pins, needles or coins are thrown into the water as the 'wish' is made. There are also many 'Haunted' Wells, this of course being the old idea of the nymph of the spring, so common in classic literature. Among the reputed haunted springs or wells are the following: Rostherne Mere (Cheshire), Croft Pasco Pool and Dozmare Pool, both in Cornwall; but these are lakes rather than wells, as is the Silent Pool in Surrey, with its ghost of the Saxon maid, Mermaid's Pool (Hayfield), St. Osyth's Well (Essex), Callow Pit (Southwood), Black Pool (Longnor), Ellesmere (Shropshire) and St. Julian's Well (Wellow, Somerset).

Healing Wells were very common, the diseases believed to be benefited including sore eyes, leprosy, skin diseases and madness. It is a curious fact that the pagans seem to have had better judgment in selecting healing wells than Holy Church! Two splendid Healing Wells, which are to-day recommended by the medical profession, were detected and used by the Romans (i.e. Bath and Buxton), while I have discovered scores of mediæval Holy Wells, believed to possess the utmost virtue during the Dark Ages, so that hordes of pilgrims thronged thither, which to-day are empty, neglected and sometimes waterless. A typical example is St. Margaret's Well at Bisney, Oxfordshire. During the last century before the Reformation it was so thronged by pilgrims that a house was built over it and kept locked, while the water was sold at a guinea a quart, an enormous sum in those days. To-day it is empty, neglected and – when I last visited it – waterless.

The most interesting Healing Well in all Britain is at Bath. The well-known legend runs as follows. A certain Prince named Bladud (he was the father of King Lear, and died 844 B.C.) was stricken with leprosy and driven from his Court. He wandered into the Forest and became a swineherd. While tending his pigs he noticed that they loved to wallow in the warm mud of a swamp, and that those suffering from skin diseases were cured thereby. He followed their example, and was himself healed. Modern historians deride the legend, but it is probable that the springs were known to the Britons before the coming of the Romans, since a spring coming up out of the earth at a temperature of 117 degrees Fahrenheit would surely attract the attention of the most ignorant savage.

A clergyman named Groves of Claverton, has satirised the neighbouring town of Bristol – always the deadly rival of Bath – in these verses :

> ' When Bladud once espied some hogs,
> Lie wallowing in the steaming bogs,
> Where issue forth those sulphurous springs
> Since honor'd by more potent kings,
> Vex'd at the brutes alone possessing
> What ought t' have been a common blessing,
> He drove them thence in mighty wrath,
> And built the mighty town of Bath.
> The hogs thus banished by their Prince
> *Have liv'd in Bristol ever since* ! '

The Roman Baths date from about A.D. 44, and can still be seen much as their builders left them. The great bath is 111 feet long and 68 wide, and was originally lined with lead a quarter of an inch thick. There are two smaller Roman baths, one circular and the other rectangular ; and very interesting inscriptions have been discovered, dating from the Roman period, which prove that Bath was even then visited for the cure by people from the Continent. A visitor from Trier (Germany) gave thanks for his cure, as did another from Metz, and a lapidarius (stone-merchant) from Chartres.

The historian Solinus, writing in the third century, says : 'In Britain there are hot springs, furnished luxuriously for human use ; over these springs Minerva presides.' Here we have the idea of the deity (female of course) of the spring. There was a native goddess in charge here before Minerva. Her name was Sul, hence the Roman name for the city, 'Aquæ Sulis.'

Another most interesting relic is the 'Bath Curse,' found scratched on a piece of metal. It was written *backwards*, just as in mediæval legends it was impious to read certain psalms backwards, the Lord's Prayer said backwards being a potent bane. The piece of metal was thrown into the water. It reads thus : ' May he who carried off Vilbia waste away like that dumb water, save only he who . . . her.' The blank is not deleted by censor, it has rusted out. What kidnapping romance of twenty centuries ago lies behind this simple inscription ? And how romantic the idea that those same waters, which healed the sick two or three thousand years ago, are still flowing steaming hot from the bowels of the earth, and are as potent for healing to-day with their radium-charged waters, as they were when poor Bladud tried them.

Another Roman Well which still does good work is at Buxton. In mediæval times it was dedicated to St. Anne. A chapel which stood over it was pulled down at the Reformation, while the Roman work was unfortunately destroyed in 1709. There is no doubt that St. Anne was substituted for the pagan divinity to whom the well was formerly sacred ; as Grimm, in his *Preface to Teutonic Mythology*, says, ' Sacred Wells and fountains were re-christened after saints, to whom their sanctity was transferred.' Curiously enough, we have two wells in England which have escaped this process and still bear the names of heathen deities, e.g. Woden's Well (Gloucestershire) and Thor's Well (Yorkshire).

A very famous and exceedingly interesting example of the Wishing Well is the celebrated Well of St. Keyne, near Looe, Cornwall. Tradition says that it was formerly overshadowed by an oak, an ash, an elm and a withy, all growing from the same trunk. These ancient trees were blown down in the great storm of November 1703, but when I visited the spot in 1927 there were two trees – an elm and an ash – growing very close together and so close to the well that there is danger they may destroy the stonework of the wellhead. Here, again, we have the usual association of the tree and the fountain ; the oak and ash are especially significant, since the former was the sacred tree of the Druids, and the latter was symbolical of universal life. The fact that these three or four trees were formerly growing so closely together, their roots so intertwined that they seemed like one tree, must have appealed to the pagans, especially because a clear spring issued beneath. As they have grown up again together after the whole was thrown down two and a quarter centuries ago, I think we may

THE BLACK PRINCE'S WELL, HARBLEDOWN, NEAR
CANTERBURY.

TAKING A DRINK AT THE WELL OF ST. KEYNE.

To face p. 92.

assume they have again and again arisen, phœnix-like, from ruin, and the present trees are the last of a very long-lived family who have occupied the spot almost from time immemorial. The legend of the Well has been amusingly rendered into verse by Southey in his famous poem:

' A well there is in the West Country;
And a clearer one never was seen;
There is not a wife in the West Country
But has heard of the Well of St. Keyne.

If the husband of this gifted well
Shall drink before his wife,
A happy man henceforth is he,
For he shall be master for life.

But if the wife should drink of it first
God help the husband then!
The stranger stooped to the Well of St. Keyne
And drank of the water again.

" You drank of the Well I warrant betimes?"
He to the Cornishman said.
But the Cornishman smiled as the stranger spake
And sheepishly shook his head.

" I hastened as soon as the Wedding was done
And left my wife in the porch.
But i' faith she had been wiser than me,
For she took a bottle to Church!"'

An interesting variation from the usual run of Wishing Wells is to be found in Denbighshire. It is the Cursing Well of St. Ælian – drop in a pin and utter the name of the one you desire blasted.

The most celebrated well in this part of the country is St. Winifred's Well at Holywell. The great festival there is on May 1st, when processions are held resembling Lourdes, and, like that resort, it is under Roman Catholic patronage. It is claimed that miraculous cures have been effected, and the place is frequented by deaf, dumb, blind and paralysed persons, some of whom have left behind crutches as symbols of their safe recovery. The well is said to have gushed forth miraculously from the spot where the severed head of St. Winifred struck the ground. She was a Welsh Princess who was courted by a prince named Caradoc; she refused to marry him as she desired to enter a convent. His rejection so angered the Prince that he pursued her and struck off her head with his sword. The maiden was restored to life by St. Bruno and became an Abbess, while the murderer fell dead.

A very remarkable instance of a pagan Holy Well under a Christian Church is to be seen in the crypt of Winchester Cathedral. It is probable that the same site had been used for Celtic worship, for Roman idols, and lastly for Christianity. This Cathedral, by the way, is the only one in the world where a diver was employed to lay the foundations, and I was fortunate enough – during the recent restorations – to see him at work. He underpinned the ancient foundations, which were sinking, with concrete.

A still more remarkable example of pagan survivals was formerly to be seen in France, namely, a large phallic emblem carved on a door of Toulouse Cathedral. It was destroyed by the population during the Revolution. Another – made of wood covered with leather – was formerly kept among the treasures of St. Eutrope's Church, Orange, and was burnt by the Huguenots. Close by was a Holy Well said to cure sterility.

The crude emblem is common in Eastern cults to-day, and is well known all over India as the emblem of Siva. Crude, and even disgusting, as the lingam appears to European eyes, it suggests nothing of this to the natives. Every day the men place yellow flowers before it, and the women pour water or melted butter over it. It is worn as an ornament, and carried to the beds of sick men, just as the Catholics take a crucifix.

When the Catholic missionaries first reached India there was a violent controversy between the Capucins and the Jesuits, as to whether Christian converts should be permitted to wear this amulet. The former forbade the practice and wished to replace it with the medallion of a saint, but the natives would have nothing of this, and protested that the Jesuits allowed the charm. Finally the Capucins gave way, and permitted the emblem of Siva to be worn, provided that a cross was carved on it! This affords an interesting example in almost modern times of how pagan emblems have been absorbed into the ritual of Christianity.

But I must return to our Holy Wells. They are still very common in England, though some of great interest have disappeared during the last twenty years. A very tragic instance is the famous 'haunted' Drumming Well at Oundle, which has now been built over; but I have met an old inhabitant who has heard the mysterious drumming which gave it its ghostly reputation. Mr. R. C. Hope, F.S.A., in his excellent work on Holy Wells, mentions about 450 in England alone. Actually the number is somewhat larger, as no writer, however industrious, could discover them all, and I have found at least a dozen of some interest which are not mentioned in his book. Needless to say, the Celtic fringe contains the most, Cornwall having 40, Northumberland 35, Cumberland 26. Compare this with Essex one, and Buckingham, Cambridge and Hertford two each.

The legends of Holy Wells are thus mainly Celtic, and they are widely found in Scotland and Ireland. Space forbids reference to the former country, but I must refer to the very remarkable survivals of paganism which are associated with the Holy Wells of Ireland.

It is well known that the south and west of Ireland have long lingered far behind other parts of the United Kingdom both in education and culture, so it is naturally in the remote and backward districts of the Emerald Isle that we seek for the most interesting remains of that pre-Christian nature worship which men call paganism.

A typical instance can be seen to-day at the

Holy Well of Doon in a remote corner of Donegal. Clustered round the well are a number of votive crutches, and these and every bush in the vicinity are swathed in scraps of rags or strings of beads. These are really votive offerings to the goddess of the well, and I have seen exactly similar rag-covered bushes in darkest Africa. The well is still resorted to, and the ritual is as follows. Walk barefoot to the well, say five Aves, five Paternosters and one creed; then drink at the well, say five Our Fathers and Five Hail Marys for the bottle of water you take away, and one for the priest who blessed the well, and another for the man who built a shelter over it.

That point about going barefoot interested me, and I have taken much trouble to discover how far this method of self-torture was carried a century earlier. Happily I discovered a well-written book published in Dublin in 1840 by Mr. P. D. Hardy, which contains a minute description of the Holy Wells of Ireland and the religious rites celebrated there. As I anticipated I found the ritual was entirely pagan, save only in name. Rags, horseshoes, pins, needles, coins, butter, etc., were thrown in the wells or hung on the trees; and the pilgrims endured cruel self-inflicted agonies. At some stations revolting orgies also occurred.

The worst of these was at St. Patrick's Purgatory, Lough Derg, which is still resorted to by modern pilgrims from June 15th to the end of August; though, of course, the worst crudities of the past have now disappeared. The 'Retreat' formerly lasted for nine days, during which the penitents were not permitted to sleep and had only one meal of bread and water in twenty-four hours. Until the year 1781 twenty-four hours were spent in a dark airless underground cavern, but so many of the fanatics lost their lives that this was closed and twelve hours in a dark chapel called 'the prison' substituted. In 1840 the 'Retreat' had been reduced to three days, but even then it was fatal to many of the devotees.

A Mr. Carleton (who braved the whole ordeal) has graphically described his experience in *The Lough Derg Pilgrim*. He tells of his long walk thither barefoot till his feet were in 'griskins,' and the torture of going over sharp rocks on bare knees, and closes with these words : 'There is not on earth ... a regulation of a religious nature more barbarous and inhuman than this. It has destroyed thousands since its establishment, has left children without parents and parents childless. But what is worse than death, by stretching the powers of human sufferance until the mind cracks under them, it is said sometimes to return these pitiable creatures maniacs.' At Holy Island was another agonising scene for which I will quote Mr. Hardy. After making some two hundred and eighty circuits of St. Mary's Church, the last round is performed on the bare knees, including a *vertical drop of over a foot*. 'This,' he says, ' is a most painful operation. The writhing postures, the intense agonies, and the lacerated knees of the votaries are most distressing to the spectators. . . . Then they must go on their bleeding knees through the rough stones in the church to the east end, when in a posture of the most profound reverence they must kiss a particular stone.' Perhaps the most interesting of all the Holy Wells of Ireland is on Croagh Patrick, and this is an obvious pagan High Place. The fine conical, or sugar-loaf, profile of the mountain – an obvious phallic symbol – had doubtless made it a place of Baal worship from the very dawn of time, and here we have all the stigmata of primæval paganism – the conical peak, the holy well and the sacred tree. Mr. Hardy also refers to obscene orgies which I will not relate, except to mention that he speaks of them as happening at many of these places of pilgrimage, an apt reminder that the pagan Earth-Mother is remembered there as well as Baal.

It is still a place of pilgrimage, but in 1840 the penance was very severe; including a long crawl over sharp rocks on bare knees till they were streaming with blood, with three nights in the open spent in prayer and fasting. The Rev. James Page (in a work published about 1830) graphically describes the terrible suffering of the pilgrims here, whose bare feet and knees were badly lacerated, so that every step left a bloodstain. He also describes the rites performed at St. Patrick's Bed by barren women desiring children. I will sum it up in his own words, ' The abominable practices committed there ought to make human nature, in its most degraded state, blush.' Though the good clergyman was obviously shocked and astonished, there is nothing here to surprise the student of history ; these are merely the rites of Baal and Baalath (the Sun-god and the Earth-Mother), and all were exactly paralleled in classical times, and still earlier in the heathen temples and high places.

Consider for a moment the graphic description in 1 Kings xviii. of the priests of Baal invoking their god. This was on a hill-top, and obviously near a well ; or where were the barrels of water obtained ?

'And they cried aloud, and cut themselves after their manner with knives and lancets till the blood gushed out upon them.'

As recently as 1925 I went to see a very holy marabout in an African village, who had a wide reputation for sanctity throughout the whole region. He was an old man, with thin worn face and grizzled hair and beard, his sightless eyes had been extinguished by his own hand, and he still tortured himself horribly for the good of his soul. It was a revolting experience. We were crowded in a small palm-thatched mud-walled hut, having a great brazier of glowing coals at one end. The 'saint' first plunged iron skewers through his cheeks, his forehead, and throat till he looked like the fretful porcupine. Then he approached the brazier and drew from it a red-hot iron, which he licked and placed in his mouth. Finally he removed his scanty garments and stood in the glow of the firelight, naked save for a loin-cloth. Then he seized a great handful of dry straw and ignited it at the brazier, and standing with the blazing mass in his bare hand, bathed his emaciated body in the flames, till the narrow room was full of the foul reek of burning flesh, and one European visitor near me fell in a dead faint. This is an instance of the pagan idea of self torture, found in Islam to-day.

THE MORRIS DANCERS

'It was my hap of late, by chance,
To meet a Country Morris Dance,
When, cheefest of them all, the Foole
Played with a ladle and a toole;
When every younger shak't his bells

. . . .

And fine Maide Marian, with her smoile,
Show'd how a rascall plaid the roile (etc.)'
                    (*Old song printed in 1614.*)

THE Merry Month of May was in mediæval times the gladdest of months, and even to-day – when modern inventions have lightened the gloom of the winter – we are all cheered by the return of the sun and the coming of bud and blossom. And as May was a festival of youth, it is also the month of flowers and garlands and dancing, especially Morris Dancing.

The Morris Dancers usually make their first appearance on May 1st, and emerge again at Whitsuntide, when the great festival at Bampton (Oxon) is held. They are to be found at country fairs and fêtes all through the summer, and one or two troupes occasionally appear at Christmas, but May is the Morris Month, and I propose to describe two of the best shows of the kind which can be seen to-day.

The Morris Men are the most picturesque of all the survivals of mediævalism which remain to us to-day. Like the Mummers – with whom

MR. WELLS AND HIS TROUPE OF MORRIS MEN.
Note the sword impaling the cake.

To face p. 102.

they are closely akin – they wear a remarkable traditional costume, and make use of beautiful old folk-songs which have been handed down by ear for centuries, and some of which have never appeared in print.

But while the Mummers are handicapped by the gloom and murk of December, the Morris Men caper on green lawns in the full glory of early summer, and I have seen few more beautiful sights than a Morris Dance in a lovely old garden, when the white costumes and gay ribbons of the dancers flit in and out between the rich greenery of fresh foliage and the brighter tints of the flower-beds. And while the bright colours and tuneful measures delight both eye and ear, the hoar antiquity of the usage makes its appeal to the mind.

What are the origins, what the meanings, of this ritual so widely dispersed, so faithfully obeyed and so long followed?

My answer is that these picturesque ceremonies have their origin in the dim past, when in the childhood of mankind, before the beginnings of organised religion, simple rites were performed to ensure the safety of the harvest, the fertility of flocks and herds, or the success of tribal hunting. The month of May is particularly rich in these vestiges of the past, because the re-birth of nature was a time which naturally appealed to primitive and uninstructed man.

Hence the prevalence in nearly or quite all pagan religions of that bi-sexual dualism – the male and female deity, identified with the sun and earth or moon – the former requiring to be propitiated by human or animal sacrifices, and the festivals of the latter associated with the removal of sexual inhibitions. To simple man the great miracle in nature was sex, by which a male and female could unite together and create a new life ; and he believed that the two great lights in the heavens were themselves the male and female deities whose union had created the universe as he knew it. Hence the supreme importance of the principle of fertilisation in all early cults and pagan religions ; the May dances and garlands, the dressing of wells, and the tree-worship of the Druids.

The tree was the noblest example of plant life, and so was worshipped by rings of dancers, and ceremonies were performed at sacred groves and wells which were believed to give renewed vitality to the spirit of fertilisation, and so help the people to obtain larger crops and more cattle.

While most of the May ceremonies which remain to-day undoubtedly relate to the fertilisation idea, there are several features in the Morris ceremonies which seem to me characteristic of hunting usages (see footnote, p. 105).

'What !' I hear some choleric M.F.H. exclaim, 'does the idiot imagine we hunt in summer ?'

As a keen follower of the hounds myself, before age intervened, I am well aware that fox-hunting is a pastime of the winter months; but the mediæval hunting man knew naught of Brer Fox. Not for him the 'pursuit of the uneatable by the unteachable'; he spurred his steed after a nobler, and more palatable, quarry – the great red deer. And we may be sure that those earlier huntsmen went after the wild boar, or timid deer, which would give them fun and food as well. And deer-hunting – as the Exmoor riders well know – is carried on in summer and early autumn, as motorists in Minehead and Porlock are aware. In Sherwood Forest the mediæval hunting season for deer was from June 24th to September 14th and November 11th to February 2nd.[1] In many of these old dances, and particularly at Padstow, a hobby-horse is a main feature in the proceedings, and surely this is a hunting allusion. And at Bampton there is a swordsman who carries a naked sword, impaling a cake, brightly beribboned; this – if I mistake not – relates to the victim of the hunt. One learned gentleman whom I encountered at Bampton told me that the cake was no doubt the Host. I do not believe it. To the Catholic mind the Host was so sacred as to be untouchable – the very Person of the Deity. Even to touch it would be to risk the doom of Uzzah, to offer it violence was unimaginable. Were not certain unfortunates put to death with horrid tortures

[1] Mr. R. J. E. Tiddy, in his book *The Mummers' Play*, states that at Leafield, in the forest of Wychwood, there was formerly a licensed deer-hunt held in connection with the Morris Dances, and that in some other Wychwood villages animals were slain in the same connection.

for having pricked it with a needle – presumably to ascertain whether it were still bread, or had changed its substance?

Some of the dances undoubtedly have to do with seed-time; the Bean Dance, with its thumping and clashing of staves, mimics the action of the dibbler who makes a hole in the soil to take the seed. Some writers derive Morris from Moorish, and the word (like the Turkey Snipe – or Turkish Knight – in the Mummers) may have come from the Crusaders; though I hope I have convinced you that its origin is far earlier. I have found references to them which suggest that they were known in the fourteenth century, but no earlier mention. Mr. Wells, the leader of the Older Bampton troupe, tells me that they have a continuous history of over five hundred years, and the Ilmington (Warwickshire) Morris Dancers have been celebrated since the fourteenth century. Unfortunately, however, they have not a continuous history, having lapsed for some time. They were revived some years ago, and are now under the leadership of that splendid old veteran Sam Bennett of Ilmington. He has been connected with the village for forty years, and has trained many Morris troupes, and I could learn more in five minutes' conversation with Mr. Bennett than in hours of study. He has a wonderful old fiddle, dating from 1640, and can play over three hundred tunes by ear. Some of these are ancient folk-music which have never been

MR. WILLIAM WELLS OF BAMPTON, OXON.
The troupe which he leads has a continuous history for 500 years.

MR. SAM BENNETT OF ILMINGTON, WARWICKSHIRE, WITH HIS FIDDLE (DATE 1640).
He can play 300 tunes by ear.

THE MORRIS DANCERS.

To face p. 106.

committed to print, and this genial old veteran can roll them off, one after the other. He wears a Shakespearean smock, over a century old, as is his wonderful beflowered hat.

Another mighty Morris Man is Mr. William Wells of Bampton, to whom I am indebted for much information. He is an enthusiast for Morris Dances and Morris Music, and has had a lot to do with the re-popularising of folk songs and dances. He tells me that it was he who furnished the tunes printed in *The Morris Book*, that monumental work in many volumes by Mr. Cecil J. Sharp – though I imagine Mr. Macilwaine had a good bit to do with it also. Certainly Mr. Wells is a mine of information and a splendid natural musician, and plays all those ancient, catchy tunes which even one as non-musical as myself soon learns to recognise.

Readers who wish to see the Morris Dancers will find that the county of Oxford is their Mecca. There are three troupes in or near the city itself. The Oxford University Morris Dancers (black shoes), the celebrated Headington Quarry Troupe (white shoes) and the Oxford City Police troupe. I have seen all of these on many occasions, and have met Mr. Kimber – the leader of the Headington men – who is worthy to be numbered with Messrs. Bennett and Wells, and so form a mighty trinity of Morris Dance Leaders. The first two troupes nearly always appear in Oxford on May morning, between 6 and 7 a.m. The performances are given in the open streets at various spots where traffic permits, and generally the following : a side street near Magdalen Bridge, Broad Street, and the open space opposite the Martyrs' Memorial. Both wear the usual white cricket shirts and trousers, with bright ribbons crossing the back and chest. The bells, as always, are stitched on pieces of coloured leather, and bound round the legs. These two troupes wear cricket caps, which is regrettable, as these are far less attractive than the beribboned and beflowered headgear of the Bampton men. Some troupes remain faithful to ancient toppers, gaily adorned, which are most suitable.

Each troupe consists of eight persons, six dancers, a musician and a fool. The latter is dressed as a clown, and has a bladder on a string attached to a stick, with which he keeps the crowd back when they press too closely on the dancers. The musician is, of course, the leader of the troupe and usually plays the fiddle, though one well-known exponent remains faithful to the concertina, and plays it exceedingly well. In earlier days the celebrated Robin Hood, with Maid Marian and other members of the outlaw band, were included. As these were very definitely huntsmen – even if illicit – this is some support to the theory that the Morris Ceremonies had to do with hunting.

I have discovered an extract from a fifteenth-century MS. which sums up in a very interesting way the round of the folklore year :

'At Ewle we wonten gambole, daunce, to carol and to sing,
To have gud spiced sewe, and roste, and plum pie for a king;
At Easter Eve, pampusses, Gangtide-Gates did oilie masses bring;
At Paske begun oure Morris, and ere Pentecoste oure May,
Tho' Roben Hood, liell John, Frier Tuck, and Mariam deftly play,
And lord and ladie gang 'til kirk with lads and lasses gay;
Fra masse and een songe sa gud cheere and glee on every green,
As save oure wakes 'twixt Eames and Sibbes, like gam was never seen
At Baptis-Day, with ale and cakes, bout bonfires neighbours stood;
At Martlemas we turn'd a crabbe, thilk told of Robin Hood.'

It will be remembered that the old ballads describe Robin Hood as the Earl of Huntingdon, and his wife – Maid Marian – was said to be Lord Fitzwalter's daughter. Their plucky resistance to the haughty Norman, and their kindness to the poor (with whom they shared the booty taken from the rich alien oppressors of the country) naturally made them heroic national figures to the common people. That is why Robin Hood so frequently appears in the early Morris Troupes, and is also sometimes found among the Mummers. Maid Marian may have been the forerunner of the May Queen, with whom she became subsequently identified ; and in Dalrymple's Extracts from *The Book of the Universal Kirk* (1576) Robin Hood is styled ' King of the May.' Latimer also mentions Robin Hood and his men, saying that Robin Hood's Day was kept by the country people in memory of the great outlaw, and at least half-a-dozen sixteenth-century writers mention Robin Hood, Little John, Maid Marian and Friar Tuck as popular personalities in the Morris Dance. Tuck was chaplain to Robin Hood. He wore the habit of the Franciscan Order, which was the only order exempt from episcopal jurisdiction, and carried the wallet of the mendicant orders of religion.

But I must return to the Bampton Morris Festival, which is the most interesting in England and quite accessible, as the little town is situated on an excellent main road midway between Farringdon and Witney. It is a region of meadowland through which the winding Thames slowly flows, of stone-built villages embowered in trees, and fine churches ; and though there are those to whom flat country has no appeal, there is to me at least a charm in its peaceful pastures roofed over by a mighty arch of sky, for here, as in Holland, we realise the fact that level country affords magnificent sky effects. It is remarkable that the guide-books (some of which devote several pages to Bampton church) have not even mentioned the fact that the little town boasts the oldest troupe of Morris Dancers in all England, and that it is *en fête* on Whit-Monday.

This is a great mistake, as the average tourist would gladly sacrifice a dozen better churches than that of Bampton to see such a festival as I am about to describe. I quickly covered the $15\frac{1}{2}$ miles from Oxford city and found the village full of children carrying May garlands, and dressed in their Sunday best. On enquiring for the Morris Men, I was directed to a private house, whose large garden was open to the world while Mr. Wells and his gifted troupe were performing.

It was a fascinating sight. All were dressed in white, with shady straw hats, both hats and garments being decorated with bright pieces of gay ribbon. There were six dancers, wearing pieces of brightly coloured leather strapped to their legs, and having little jingling bells sewn on.

In addition there was the fiddler or leader of the troupe (Mr. Wells) and the fool – or comic – complete with bladder on stick.

The dances are performed with rare skill and have a distinctive character of their own. There is no languor about the Morris Dancers, they do not sway or glide. They stamp vigorously and kick as if they meant it. The bells are expected to ring loudly, and it takes a real stamp to do this ; and the foot when lifted is never drawn back, but always thrust forward. The forward or stepping foot is lifted as in walking, and then straightened to a kick to make the bells ring, while at the same instant a hop is made on the rear foot. The dancer alights on the toe to break the jar, but almost immediately allows the heel to fall so that he stands on the flat foot. The most picturesque of all the dances in my opinion is the Bean Dance, with its clashing and thumping of staves; the six men standing in threes facing each other. The movements undoubtedly represent the setting of seed.

Very graceful is the Handkerchief Dance, which I have found most difficult to photograph, because the waving white of the large handkerchiefs blurs the picture save at high shutter speeds, and these would mean under-exposed pictures. In part of this dance the six men face each other with a white kerchief in each hand, held by the corners, and wave them up and down, first extending the hands above the head, and finishing behind the back. It reminds me slightly of the handkerchief dance which I have seen performed by the Ouled Naïls of the Sahara Desert.

The Pipe Dance requires much skill and agility. Two long clay pipes are placed crosswise on the ground, much as claymores are for the Highland Fling, and the dancer threads his intricate steps about them without ever breaking one. This, of course, is for one performer only, as is the Broom Dance, in which an ordinary household broom is held in the right hand. I have already mentioned that the Morris Dancers consist of men only, but there is one – 'The Princess Royal' – in which there is one lady and one man. To those of my readers who wish to know all about

MR. WELLS CONDUCTS HIS TROUPE.

To face p. 112.

the intricate steps of the numerous Morris Dances (and there are at least a hundred of these), I would recommend *The Morris Book*, by Messrs. Sharp and Macilwaine. This work also includes the music of many of the delightful little tunes, with which I am familiar from seeing and hearing the Morris Men, and which have a real charm of their own.

And indeed I know of few more charming sights than that of a group of Morris Men performing their complicated evolutions, jigging and capering, now advancing towards each other, now retiring, following in circles, or advancing in double file over some green lawn, while the rich foliage of an English summer forms a delightful background to their gaily coloured figures. Nor must it be forgotten that it is a labour of love to them to keep alive this picturesque relic of mediævalism, and to rescue from the oblivion, which would have otherwise overwhelmed it, a really delightful fragment of earlier days.

Though the garden dances are very pretty, the most interesting item in the day's work was the united festival in the Market Place. Here the two troupes joined forces, and under the combined leadership of two veterans of the craft – Messrs. Sam Bennett and William Wells – performed some graceful and intricate dances with rare skill.

And it speaks volumes for the ignorance or indifference of our countrymen to this remarkable free show, that there were barely a score of strangers present to see it, and most of them were undoubtedly passing motorists who had come upon it by accident.

If this book is able to introduce a wider public to the many unknown charms and hidden delights of our fair countryside, it will not have been written in vain.

THE FESTIVAL OF THE DUNMOW FLITCH

NOW a Bank Holiday attraction, usually on Whit-Monday, this interesting event does not take place at Great Dunmow, which boasts Dunmow Station on the L.N.E.R., but at the village of Little Dunmow, about three-quarters of a mile from Felstead station. London motorists can easily reach it in less than two hours. It is 37 miles from the G.P.O. to Dunmow via Chigwell and Chipping Ongar. In Great Dunmow, enquire the route to Little Dunmow; it is only about three miles, but slightly intricate.

The picturesque custom of the Dunmow Flitch is of very ancient origin. Robert Fitzwalter in the reign of King Henry III (1216–72) instituted the arrangement by which any *man* who had not repented of his marriage, sleeping or waking, for a year and a day, might go to Dunmow and claim a flitch of bacon. It is interesting to notice that from its inception in the thirteenth century right up to the eighteenth century the claim was purely a one-man job. In the view of our ancestors a happy marriage was one that was satisfactory to the man. The double claim of a married couple is not heard of before the eighteenth century. The applicant had to take oath before the prior, convent and townsfolk, kneeling on two hard pointed stones in the churchyard, and was then carried in an ancient wooden chair through the town. The Priory was destroyed at the Dissolution, except a part of the church, which is now the parish church of the village, and the historic chair dating from the thirteenth century is preserved in the chancel. There are also alabaster effigies of Walter and Matilda Fitzwalter (1198), not, I fancy, the founder of the custom, but probably his son and daughter-in-law.

Dugdale gives us the names of three claimants for the Flitch in pre-Reformation times. In the seventh year of King Edward IV a gammon of bacon was delivered to one Steven Samuel of Little Ayston; in the twenty-third year of King Henry VI a flitch was delivered to Richard Wright of Badbourge near Norwich; and in 1510 a gammon was given to Thomas Ley, fuller of Coggleshall.

In 1701 we have a record of two gammons being delivered to John Reynolds and Ann his wife, of Hatfield Regis, who had been married for ten years; and two more to William Parsley of Much Eyston, butcher, and Jane his wife, who had been married for three years. The jury consisted of five spinsters, and the following metrical oath was imposed on the claimants:

' You shall swear by custom of confession,
  If ever you made nuptial transgression,
  Be you either married man or wife,
  If you have brawls or contentious strife
  Or otherwise at bed or at board,
  Offended each other in deed or word,
  Or since the Parish Clerk said amen,
  You wish'd yourselves unmarried again,

> Or, in a twelvemonth and a day,
> Repented not in thought any way,
> But continued true in thought and desire
> As when you join'd hands in the quire.
> If to these conditions without all feare
> Of your own accord you will freely sweare,
> A whole gammon of bacon you shall receive,
> And bear it hence with love and good leave,
> For this is our custom at Dunmow well knowne,
> Though the pleasure be ours, the bacon's your own.

No doubt at one time the custom was observed with the utmost seriousness, since we only find three successful claimants in as many centuries, and a flitch, or even gammon of bacon, was a handsome prize to the poor.

To-day the thing has degenerated into an annual Bank Holiday revel, in which the main object is to provide good fun for the crowd. Proceedings are run as a kind of parody on the Divorce Court, with Judge and Counsel in imposing robes of office. Much fun is caused by the cross-examination of the claimants, and though the whole is excellent fooling, it has lost most of its interest to the serious student of folklore. This interesting old custom was not peculiar to Dunmow, but there are traces of it in Brittany, and also at Wichenovre Manor. In recent times the Bank Holiday revel idea has been extended further, and a 'Stonehenge Flitch' is now given every year at the Amesbury (Wilts) Carnival, with similar ceremonial.

WHIT-MONDAY AT SOUTH HARTING, SUSSEX

SOUTH HARTING is a pleasant village on the Hants border, about five miles from Petersfield, and well known to south country motorists on account of its steep hill, which was formerly much in vogue for hill climbs.

The district is replete with folklore, and quaint traditions are still recalled by the old folk of Harting's famous witch, one yclept Mother Digby, who lived in Hog's Lane. She was said to roam the country in the form of a hare. Proof positive was adduced by Squire Russell, who always lost his hare down a drain near her house. One day the dogs seized the hare by its hinder parts, but it escaped them down the drain, and the Squire, instantly opening the Witch's door, discovered the hag rubbing the portion of her anatomy corresponding to that which had been seized by the dogs. The Old Club has a festival on Whit-Monday, when members attend the church carrying carved hazel wands, and plant beech boughs. This may be a reference to the staves carried by the Canterbury Pilgrims, some of whom may have passed through the village, though it is certainly not on the direct route of the 'Pilgrim's Way.'

May or *June*

THE KNUTSFORD FESTIVAL, CHESHIRE

KNUTSFORD is a very ancient market town in Cheshire, supposed to have taken its name from King Canute (or Knut), who crossed the stream here after a battle. The town is of great interest to the student of folklore, as it treasures a number of quaint survivals. The most interesting and remarkable of these is 'Sanding,' in which pretty designs are made on the streets or doorsteps by means of coloured sand. The grains are passed through a funnel and really beautiful effects are obtained, something like those ornaments which are sold at Alum Chine (Isle of Wight) in glass vessels. The custom was — and continues — in vogue at weddings, the front of the bride's house being decorated with hearts and lovers' knots, with a motto thus :

' Long may they live, and happy may they be ;
Blest with content, and from misfortune free.'

Mr. Green, the historian of Knutsford, suggests that the origin of the custom may have been a pious intention to remind us that in the midst of our greatest joy we are but dust and mortal. The Egyptians had a gruesome practice of placing a coffin in a conspicuous place at banquets, and in classical times a skull was sometimes exhibited at a feast with similar kill-joy intentions. Whatever its origin, 'Sanding' is now a quaint and beautiful idea, and shows another example of local craftsmanship, like the Rush-Bearing at Grasmere and the flower-petal pictures of Tissington. Another old verse refers to it thus :

' Then the lads and lasses, their tundishes handing,
  Before all the doors for a wedding were sanding,
  I asked Nan to wed, and she answered with ease,
  " You may sand for my wedding whenever you please." '

Knutsford is also celebrated for its great May Day festival, which is held in June, and includes a magnificent procession in which seven or eight hundred characters appear. First come the Morris Dancers, who are similar to those fully described and illustrated earlier. Then comes the very ancient and interesting character of Jack in the Green. This is a man so enveloped in boughs and twigs of greenery that only his eyes are visible, and in former days the sweeps were wont so to attire themselves. The origin of this custom may derive from the remotest antiquity ; a man enveloped in a tree certainly suggests some form of nature or tree worship. Thus the Druids were tree worshippers, and, as I have pointed out on page 85, tree worship was once very common. There was also a barbarous custom in Druid times of enveloping victims in cages of osier or other twigs, and setting them on fire. One cannot help thinking that the Jack in the Green is a vestige of this old idea.

After the Jack in the Green come the Jesters, the May-pole Dancers, and then Knutsford's famous Sedan Chair. Knutsford does not forget that Mrs. Gaskell lived here, and was buried in the grounds of the Unitarian Chapel. The Sedan Chair was used by local ladies in *Cranford* days, and is a prized relic.

Next comes the Queen of the May, and though her crown be but tinsel she has radiant youth and beauty that many a real queen would give half her realm to possess. She is escorted by a Sword-bearer, Beefeaters, and footguards, and attended by ladies in waiting, train-bearers and maids of honour. After parading the streets the procession reaches Knutsford Heath, where the beautiful ceremony of Crowning takes place. This is one of the most interesting festivals in England and attracts thousands of spectators.

I am indebted to Mr. Henry Walker of Gaston-le-Marsh, Alford (Lincs.) for some of the foregoing information. In reply to my enquiry Mr. Walker informs me that though it is known as the May Day Festival (and partakes of the usual character of such events), it has no fixed date, but is arranged every year. In 1929 the date was June 14th, being so arranged to suit the convenience of Princess Mary, who wished to attend. Readers wishing to see it should write beforehand to Hon. Sec., May Day Fête, Knutsford.

JUNE

RIDING THE MARCHES AT HAWICK, SCOTLAND

THE TOWN CRIERS' CONTEST AT PEWSEY, WILTSHIRE

MIDSUMMER MORNING AT STONEHENGE

THE DRUIDS' FESTIVAL AT STONEHENGE

June 5th, 6th and 7th

RIDING THE MARCHES AT HAWICK, SCOTLAND

THE custom of patrolling the burgh boundaries is a Scottish equivalent to our old English custom of 'Beating the Bounds' and still exists as an annual ceremony in a number of Scottish towns. The 'Common Riding' at Hawick, or 'Riding the Marches' as it is usually known, is of more than local interest and has something of the character of a great day of national rejoicing. There is an historic reason for this. The men of the Border were familiar for centuries with that grim trinity which our English Litany styles 'battle, murder and sudden death,' with raids and counter-attacks, burnings, lootings, rape and destruction that were the common lot of the dwellers on both sides of the Border, as the stout Peel Towers still remind us. But the tragedy of Flodden field had bitten deeply into the hearts of the Scots. Like all other Border Towns, Hawick had sent the best of its man-power to the fatal field under their overlord, Douglas of Drumlanrig. All their gallantry, however, was vain. Douglas was slain with thousands of his countrymen, and so depleted was the man-power of desolated Hawick that it was twice partly destroyed after Flodden's fatal field, from sheer lack of warriors to defend it. There was scarcely a household which had not lost a

member, hardly a heart that was not bereaved.
As the poet has said :

> 'Round about their gallant king
> For countrie and for crown,
> Stude the dauntless Border ring,
> Till the last was hackit doun.
>
> I blame na what has been –
> They maun fa' that canna flee –
> But oh to see what I hae seen,
> To see what I now see !
>   Oh, Flodden Field ! '

This post-war generation, which has known long months of disappointment, anxiety and defeat, can understand the joy of the men of Hawick when an unexpected victory gladdened their sad hearts.

Riding the Marches, then, is a triumphal procession to commemorate a victory, and a victory snatched in the darkest hour of defeat. It was in 1514, soon after Flodden, that news reached Hawick that a body of English raiders had entered the district and were camped at Hornshole, some $2\frac{1}{2}$ miles down the Teviot, where the river narrows into a deep dark pool. The 'callants,' or young men of the place, rose early, stole silently to the spot, and surprised the enemy asleep. Many were killed before they knew what was happening and few escaped. The lads returned in triumph to their homes, laden with spoils.

So now we have a triumphant procession, with all the pomp and panoply of war, in which the Cornet leads his callants again, with a blue flag – exact copy of the original used on that glorious day – fluttering in the breeze. The pipers go ahead with their martial music; there are madly executed 'reels' and 'Highland Flings,' and much singing of an old ballad.

Those who desire more information on this old custom are referred to the book entitled *The Hawick Tradition of 1514*, published by Craig and Laing.

Since writing these lines I have had an opportunity of visiting Hawick and seeing the latest of its 'Common Riding' festivities. The proceedings start on Thursday evening and continue till Sunday morning, and always take place between June 5th and 11th. The shops close and the whole town gives itself up to the festivities.

While the celebrations are largely in commemoration of the local victory over English troops after Flodden, they include most interesting survivals of pagan practices. We find this in the ancient slogan, 'Teribus ye Teri-Odin,'[1] in which two Norse deities are invoked; and still more in the procession at sunrise to the Moat, which resembles and recalls the Druid festival at Stonehenge and other relics of Sun-worship. It should be noted that June is the period of sun-worship and of the Beltane fires.

[1] 'Teri-Odin' invokes two Norse deities, Odin or Woden, Father of the Gods, and Tyr or Ti. The gods or heroes of Norse mythology were claimed as ancestors by many early chieftains. Thus the Royal family of Norway claimed descent from Frey, and several Royal families, both English and Northern, are described as 'Sons of Odin.'

The time-table runs thus – I omit minor ceremonies which take place before Thursday evening. On Thursday evening is the Colour Bussing. The flag, which is said to be the actual colour captured at the Hawick victory (1514)[1], is brought to the Town Hall by the Cornet's lass (i.e. best girl – Cornets are unmarried) and presented to the Provost, who hands it to the Cornet, charges him with certain duties and tells him to return the flag later. Meanwhile it figures in all the processions. It has a blue ground and gold cross, and all Hawick ribbons, etc., adopt blue and yellow. Next (Friday) morning there is the Cornet's breakfast at the Tower Hotel, Hawick, after which the old song is sung outside.

The Hawick songs are of much interest, both words and music being locally composed. There is a whole book of songs – too long to reproduce here – but the most famous of them was written by J. G. Kennedy of Hawick in 1837, and has thirty-seven verses. I give the first verse and the chorus:

> 'Sons of heroes slain at Flodden !
> Met to ride and trace our common ;
> Oral fame tells how we got it,
> Hear a native muse relate it.'

Chorus :
> 'Teribus ye Teri-Odin,
> Sons of heroes slain at Flodden,
> Imitating Border bowmen
> Aye defend your rights and common.'

[1] It is really a copy, being far too new to be the original.

The next scene is a procession headed by the Cornet on horseback, carrying the flag, with ex-Cornets of the two previous years on his right and left. There are various ceremonies with the flag and a repast of curds and cream, followed by sports and races. In the evening, there is the Cornet's Ball, where all the old songs and tunes are heard and many Scotch reels are danced.

At 4 a.m. the company proceeds to the Moat, where the ancient hymn is sung at sunrise.

At 9.45 a.m. there is another procession round the town and up to the Common, followed by races and sports; and the whole affair finishes with a dinner that night which, I believe, lasts into the wee sma' hours.

It should be noted that the Common (whose bounds are inspected in the Common-Riding ceremony) was actually given to the town in recognition of the bravery of the Hawick callants (i.e. youths), so that they are carrying the captured flag, and inspecting the Common which their ancestors won.

The whole celebration is carried through with the utmost enthusiasm and attracts large crowds.

Common Riding, or surveying the boundaries, is found elsewhere in the border country, e.g. Selkirk, and other places in Scotland. There is a big affair at Linlithgow in June, which is sometimes supported by eighteen bands.[1]

---

[1] I am indebted to Councillor Richard Laidlaw of Hawick, an ex-Cornet, for much of my information.

THE TOWN CRIERS' CONTEST AT PEWSEY, WILTSHIRE

PEWSEY is a quiet little market town, little more than a large village at the head of the rich vale of the same name, standing about midway between the market towns of Marlborough and Devizes in Wilts.

Its interest to us is afforded by a remarkable festival held here every year in the month of June – the Town Criers' Competition. In spite of modern inventions, newspapers, placards on hoardings and broadcasting, the ancient office of Town Crier is by no means extinct, and several cities are exceedingly proud of their official and provide him with a uniform resembling that of a Field-Marshal.

They all attend at Pewsey, with their robes of office, and hand-bell, and competitions are held for the loudest and clearest voice and for the smartest uniform. Some wear glossy silk hats, and others magnificent triangular headgear, slashed with gold lace, something like the Lord Mayor's footman. Though the festival is little known outside the district, it is well worth visiting. I have met a number of these Town Criers, and have found them full of interesting information.

MIDSUMMER MORNING AT STONEHENGE: THE OLDEST PILGRIMAGE IN BRITAIN

' O'er his own regions lingering, loves to shine,
Though there his altars are no more divine.'

THERE is no spot in all Britain, and but few in the whole world, so charged with mystery, so redolent with romance, as that stark group of stones on the bare hillside which men call Stonehenge. For this wondrous temple is supremely the ' Riddle of the Ages,' and has long ago captured the imagination of writers and thinkers. Volumes have been written about it, learned antiquarians and astronomers have disputed and raged violently among themselves as to its origin, its age and its purpose. Even to-day we have no certain answer, but we can give what is the probable reply. Thus we know that it was not built by the Druids, but many centuries before their time; it was almost certainly a Sun-temple and was doubtless used as such by the Druids. We know from history that the Druids indulged in human sacrifices, and we have every reason to think that the earlier races who preceded them did likewise, so it is highly probable that the persistent legends of the use of Stonehenge for human sacrifices are founded on fact. For if, as seems almost certain, Stonehenge was a Sun-temple, it was devoted to a form of worship which we find all the world over, and which is almost always associated at some period

or other with human sacrifice. We find this in all
ages and both hemispheres. It was found at
Carthage in antiquity, and in Mexico in the fifteenth century. As Sir Norman Lockyer has
pointed out, not only is the axis of Stonehenge
oriented so that it points to sunrise on the longest
day, but the same feature is found in most stone
circles in Western Europe, and the same characteristic is noticeable at the very ancient holy well
at Glastonbury, which mediævalism entitled the
Chalice Well.

Sir Norman Lockyer further elaborated his
theory in a very interesting way. Taking as his
starting-point that when the temple was built it
was so placed that the sun's disc on the longest
day rose exactly over the Friar's Heel (or pointer
stone) as seen from the Altar Stone, he calculated
the amount of variation due to the known southward trend of the sun's 'furthest north' to give the
amount of error which now exists ; and from his
data stated that the temple was erected about
1680 B.C., with a possible error of two hundred
years each way. Professor Gowland, studying
the subject from the archæological side, calculated
that the period was about 1800 B.C. ; and it may
be stated that modern opinion accepts a date
near to this as being correct.

Tradition has always insisted that Midsummer
morning at Stonehenge was the occasion of a
great Sun-worship festival, characterised by
human sacrifices, and it is interesting to mention
that for many years it has been the custom of

large numbers of people to visit the temple on the morning of June 21st and wait for the dawn.

Those who are fortunate enough to arrive in time to obtain a position near the 'Holy of Holies' are favoured on clear mornings with a sight of such magnificence as cannot readily be forgotten. This is the so-called 'Greater Phenomenon,' or clear sunrise, with the orb perched on the tip of the Friar's Heel, while the Slaughter Stone appears bathed in blood. I have made several visits, but have not seen the Greater Phenomenon, which I think has only occurred about twice in the last thirty years, but I have seen the 'Lesser Phenomenon,' as the partly obscured, but still visible, disc is called. Even this is usually seen only once in three years, but all this uncertainty does not prevent a large crowd visiting the place, and sometimes there are so many – three or four thousand, I should think – as seriously to hamper each other.

I have seen few more impressive sights than this vast crowd silently waiting in the twilight for the rising of the sun. And as I stood there on the close-cropped turf and watched streamers of light flash across the sky, I could not help thinking of the scene which must have been enacted on this very spot many times through the centuries.

We will suppose with Mr. Wells that all the country was wooded then and the old temple lay half hidden in the depths of the forest. Hither before the dawn comes a vast procession led by torch-bearers and followed by priests and men

STONEHENGE FROM THE SLAUGHTER STONE.

SUNRISE AT STONEHENGE.
MIDSUMMER MORNING AT STONEHENGE.

beating drums, and a multitude of the common people scared half out of their wits. The wild beasts slink snarling into the thicket at the glare of the torches, and presently the whole multitude is grouped around the ancient stones. Gradually the sky brightens, and in the dim light of dawn the white-robed priests are seen binding the victim and placing him on the Slaughter Stone. Glorious streamers of crimson and gold flash across the eastern sky, and a vast choir concealed within the temple enclosure takes up some barbaric chant.

All eyes are turned to the Hele-Stone standing black against the glory of the dawn. Suddenly the radiant disc of the risen sun appears above the mighty stone pillar and casts its rays on the prostrate body of the victim. Then the sacrificial knife smites home, and the death-shriek of the martyr is drowned in the clash of cymbals which greets the coming of the Sun-god. It is probable that something like this really happened, and the name Friar's Heel is very suggestive. Many writers derive it from Anglo-Saxon 'Helan,' to hide, because it hides the sunrise on the longest day. I cannot help thinking that there may be another explanation. As I have pointed out on page 86, an upright pillar was the common phallic emblem of the Sun-god, and as such is found all over the world. Why should Hele-stone not be Sun-stone (from the Greek 'Helios')? Such uprights were generally placed on a hill-top, and though Stonehenge is hardly on a hill-top, it is on a hillside, as those who have bicycled thither from Amesbury are aware. We have another phallic symbol on a hill-top in Dorsetshire. It is the Cerne Giant, generally accepted as of Celtic origin, and almost certainly a god. If examined closely it is difficult to trace the outline, but as seen from the Yeovil road just outside the village, the general outline and the crude phallic symbolism are unmistakable.

It is interesting to mention that Stonehenge was a kind of 'Charing Cross' of the Neolithic period. No fewer than seven ranges of green hills run towards it, each carrying one or more of those 'Green Roads' which were the earliest highways in our island, and of which Mr. Hippisley Cox has written at length. This suggests strongly that the country round Stonehenge and Avebury was the capital and centre of the life of that period; while Stonehenge may well have been the 'Canterbury Cathedral' of religious life.

From the north-north-east come the Cotswolds, bringing men from Bredon Hill near Tewkesbury, under which woad still grows. The Chilterns and their associate hills bring the Ridgeway (or Icknield Way) from East Anglia. Due east the Old Tin Road runs along the Downs to the coast of Kent, part of which centuries later became known by the romantic title of the 'Pilgrim's Way' because it was used by pilgrims *en route* to Canterbury. This is one of the most fascinating of all our old roads; part of it in Hampshire is known as the Harroway (Hoar, or ancient, way) and it is said that along its route ingots of tin have been dug up which had been buried thousands of years ago by Phœnicians who were attacked by robbers when on their way to the Kent Coast, and being slain by the bandits, were unable to recover their treasure. Mr. Kipling refers to it in these words:

> 'There runs a road by Merrow Down,
> A grassy track to-day it is,
> An hour out of Guildford Town,
> Above the River Wey it is.
>
> Here, when they heard the horse-bells ring,
> The ancient Britons dressed and rode
> To watch the dark Phœnicians bring
> Their goods along the Western Road.'

We can safely say that this is the oldest export trade route in Britain, and perhaps in the Empire; and it not only ran eastwards from near Stonehenge to the Kentish Coast, but also continued westwards to Marazion in Cornwall. Another ancient way proceeds from the neighbourhood of Salisbury Plain by the Wealden ridge to the coast of Sussex. It was known as the Lun Way, and traces of it can be found in place names, i.e. the Lunway's Inn near Winchester, where the old trackway crosses the Winchester-Basingstoke-London highway. Still another ancient road ran across the Dorset downs, reaching the coast near the Devon border; and another proceeded via the Mendips to the Severn sea.

These romantic trackways are Britain's oldest roads, and there is reason to think that some at least were in use two thousand years before Christ. Our ancestors chose the ridges of the hills for two reasons, speed and safety. The hilltop route was straight and easy, when all the valleys were choked with forests and swamps; and it was safe because there was little risk of ambush by savage men or attack by wild beasts.

THE DRUIDS' FESTIVAL AT STONEHENGE: GOLDEN DAWN RITUAL

THE Druids' Festival at Stonehenge is one of the most interesting and picturesque of all those described in this volume, and takes place within the mysterious circle of Stonehenge which may well be styled 'The Cradle of British Religion.'

I have several times attended the festival, and am also much indebted to the Chief Druid, Dr. MacGregor Reid, for information by letter, and to the Tribune Publishing Company for permission to quote from their literature. The Druids are a religious body, who style themselves An Druidh Uileach Braithrachas or The Church of the Universal Bond, and they claim that the order has never ceased to use the temple for worship since those far off pre-Roman times when it was the metropolitan shrine of all Britain.

This claim, I think, cannot be substantiated. The temple was probably used for Sun-worship from its erection, *circa* 1800 to 2000 B.C., right down to the coming of the Romans. It is quite possible that the observance continued during the Roman occupation – but shorn of human sacrifice – since the Romans were exceedingly tolerant to other religions. It is true that the Legions ruthlessly slaughtered the Druids at Anglesea, but there were special reasons in that case for the severity.

If, then, the temple was still occupied during the Roman occupation, this would give a continuous usage of nearly two and a half millenniums, which makes the record of Canterbury Cathedral seem a thing of yesterday. It may have continued under the pagan Saxons until the growing power of the Catholic Church intervened. This we may be sure — that no Druid worship could possibly have continued during the ages of Catholic domination.

I have placed this information before the Secretary of the Druid publication department (The Tribune Publishing Co.), and he has very kindly given me the following information, which I quote verbatim:

'The records of the Hæmus Lodge of A.D.U.B. go back to the fourteenth century. Instruction is that the Order has never ceased to exist since its formation. Dr. Copee and John Harrington in Stuart times assert existence of movement as a secret order during the Dark Ages by reason of persecution by the Catholic Church. The movement in Brittany and Southern France suffered greatly from Church persecution, the Camisard movement was an outbirth of the Druid Order.

'The Shrine of the Order, which is in South London Temple (57 Cavendish Road, S.W.12) bears date 1643. The list of chosen Chiefs is complete for 500 years.'

Whatever may be the real facts as to the

STONEHENGE.

THE PROCESSION FROM THE HELE-STONE.  THE ENTRANCE AT DAWN.

To face p. 138.

antiquity of the custom, it is certain that for many years the Druids have continued their Summer Solstice Service in the temple, and also till recently were wont to deposit a small portion of the ashes of their dead within the sacred circle. For many years it was the custom to hold the service at dawn on June 21st, but owing to the increasing crowds who now flock thither by motor-car and motor-coach, the Druid Service is held a day later to avoid the crush. There is something essentially uplifting about the hour of sunrise. When I first knew Stonehenge, before the age of motors, and long before aerodromes existed, the ancient temple stood still and stark on the lonely hillside – not a house, not a tree, not a human creature was in sight. Then at the hour of dawn, when the sun rose majestically over the Hele-Stone, the effect was magical, unforgettable. And even to-day – though I do not accept the Druid theology – I am impressed by the solemn beauty of the service.

Attired in flowing white robes and scarlet hoods the Druids entered the sacred circle and formed up round the Altar Stone in the 'Holy of Holies.' On the Stone was laid the 'Sacrament,' surrounded by seven candles. The opening prayer is full of beauty; I quote a brief extract :

'God, our All Father, permanent amid all change art Thou. Thou art as Thou hast ever been, and as Thou art, so shalt Thou ever be.

'. . . We seek and find in Thee the glory of the Dawn. We seek and find Thee when the darkness of the night has fled. The sleep of faith has ever led through night to dawn. The sleep of faith will ever lead from death to life. . . . Inspire us with Thy holy will within the beauty of the morn, and send us forth to bless and purify all who would find Thee in the glory of the Dawn. Amen.'

Their hymns too are finely worded, and suit the occasion.

'God has spoken, Light arises from the depths of deepest gloom!
God has spoken, Time beginneth where but chaos once had room.
Light ariseth, Time beginneth, in God's image man is made;
Beauty reigneth, gladness liveth, all is good that is displayed.'

As the service proceeded, bells were tinkled at intervals, and the General Invocation was pronounced. Meanwhile the eastern horizon was brightening from pearl to crimson, and crimson to pink, while golden streamers were flashing across the sky. Presently the golden disc of the Sun appeared above the horizon, and the Chief Druid raised a silver cup towards it, consecrated the 'Sacrament' and drank. One by one the Brethren followed, each poured out his portion of liquid into a small glass, then rising, slowly turned towards the east, and raised the vessel as if in salute.¹

I have questioned the Druids as to the nature of their belief. It will be clear from the foregoing description that it is wholly different from the bloodthirsty rites which were long celebrated here, and would appear to embrace some of the ideals and aspirations of Christianity. They tell me that their faith can be briefly set out in the words of one of their hymns, a verse of which I quote.

> 'In life eternal we believe,
> In hope supernal we conceive
> That path that leads to highest love,
> And by that path we pass above.'

But though the subject is interesting, I must not linger here. The Druids have their place in this work, not because of their theology, but because of the antiquity of their worship,² and its real or imaginary connection with the far-distant Sun-worship, which was once observed on this – perhaps the most romantic site in the British Empire. I will therefore make some further quotations from the *Druid Journal* and other

---

[1] At the last occasion on June 22, 1930, the procedure was altered. There was a service at dawn (actually later – 5 a.m.) and another at 11 a.m. when the 'Sacrament' was observed. There were no candles but a censer of incense instead. The rites closely resemble Christian services, even to the use of bread as well as wine in the 'Sacrament.'

[2] The antiquity, of course, depends on whether Druid historical claims are reliable. But even if baseless, they deserve a place in this work by reason of so picturesque a ceremony on so historic a site.

literature of their order. To prevent misunderstanding I would repeat that these statements represent the claims of the Druids, and I do not accept responsibility for them. They are, however, full of interest in showing the real or fancied connection of Druidism with other cults. It is also, of course, true, as I have shown fully on page 85, that Sun-worship in various forms was a leading feature in that primitive nature worship which was perhaps the parent of all religions, and traces of which can be discovered in nearly every cult.

'The faith of our order is as it was in the days of Pelagius, the Teacher of Bangor, and of the brothers of Iona, Glastonbury and Calne. The murder of the Chief Druid and his followers at Calne by the adherents of St. Dunstan did not stay the life and work of the Celtic Church, which has been preserved with An Druidh Uileach Braithreachas to this day.'[1]

They claim a number of historic characters as having been members of their order, or to have held their views; among them Amen Hetep, the Pharaoh who called himself the 'Exalter of

---

[1] This refers to the alleged miracle at the Conference of Calne in 978. This was called by Dunstan, who wished to abolish the existing right of the clergy to marry. The conference was almost equally divided, and Dunstan would not have succeeded in his enterprise except for the miracle. Dunstan called on God to support him, and immediately the flooring beneath his opponents gave way, and many were killed in the ruins, while the saint and his followers were unhurt. As a result of this, the marriage of the clergy was prohibited. The killed and injured were, of course, clergy of the British Church. I find no evidence that they were Druids, but the Druids seem to regard Celtic Christians as Druids.

Aten' (or God), and whose magnificent hymn to the sun is well known by all students of comparative religion.

'The Druid Order has three divisions: Ovates, Bards and Druids. The Druid Planetary guide is Sadi-Saturn-Sabbaoth. The Druid Celestial House is Boodes, and the bright star (Arcturus) is called the Star of the King (i.e. Arthur).'

Those who are interested in Druid literature will find in it a great mass of facts, which show its connection with that primitive nature worship to which I have already referred.

In fairness to the Druids, I ought perhaps to state that they insist that they worship God the All Father, and not the Sun, and they deny that Stonehenge was ever the scene of human sacrifices. In this latter respect, however, though there is no absolute certainty, I am bound to say I am against them. It seems almost certain that Stonehenge was a Sun-temple, and as such was connected with human sacrifices. These were very often found in association with Sun-worship, as in Asia, Africa, Mexico and Peru.

One admission I must, however, make, and that is that I confirm and support the Druid Order in their protests against the enclosure of Stonehenge. It is true that to-day it is national property – through the generosity of Sir George Chubb – but less than a quarter of a century ago this unique national monument was publicly offered for sale, and there was even the possibility it might have been bought by an American and carted away. We have bitter need in Britain for a law to protect the treasures of art and antiquity such as prevails in Italy and Greece, by which export is forbidden. You can buy the thing, but must not take it out of the country. And there seems something strange in the fact that Stonehenge has somehow become private property. Here was a great national temple – perhaps the Canterbury Cathedral of the Neolithic period – which had drawn vast throngs of worshippers from far and near. As my friend Mr. E. J. Burrow, F.R.G.S., has pointed out in his book on Stonehenge, no less than seven or eight of the ancient green roads (or ridgeways which carried the traffic of England for centuries before the Roman occupation) meet near here. This clearly suggests a great metropolitan shrine, which was national in its inception, its purpose, and its usage for centuries. And only the generosity of a private individual has safeguarded for the nation one of its choicest treasures. It is good news that the horrible Air Force buildings will soon be removed, but how fatuous the military mind, which, with hundreds of square miles of open plain to choose from, must dump this unsightly horror alongside the unique Sun-temple of Western Europe! Its appeal rested on its remoteness on a bare hill-side, but that, alas, is gone for ever; and with the dreadful barbed wire all round more too of its subtle

THE PROCESSION ROUND THE 'SLAUGHTER STONE.'

ENTERING THE TEMPLE.
STONEHENGE.

To face p. 144.

charm has vanished. But even the motor-buses and café close alongside cannot entirely destroy the wonder and appeal of this 'Temple of the Dawn.'

JULY

**THE HISTORIC TYNWALD CEREMONY
THE VINTNERS' COMPANY**

July 5th

THE HISTORIC TYNWALD CEREMONY: ST. JOHN'S, ISLE OF MAN

THE Tynwald Hill is an artificial mound standing in an open space at St. John's, Isle of Man. In form it is a truncated cone, with a basic circumference of 240 feet, and a height of 12 feet, having four circular platforms at 3-foot intervals. The Isle of Man was once the property of the Vikings, and they established here their custom of holding an open-air court for the settling of disputes and the passing of laws; and to this day no law in the Island is valid until it has been pronounced on Tynwald Hill. It is well known that the Scandinavian peoples were pioneers of democratic government, in fact the little Iceland Parliament which celebrates its millenary this year (1930) is the oldest in the world. Here at Tynwald we have the only remaining 'Thing' or outdoor parliament, and the ceremony, which is most impressive, attracts thousands of visitors. Prayers having been said in St. John's Church, the Lieutenant Governor (as representative of the King), in regal uniform, preceeded by his sword-bearer and followed by the chief officers of Church and State, passes through lines of guards to the 'Hill.' Here he mounts to the topmost platform, where he sits on a crimson velvet chair, looking eastwards, and with the sword of State laid before him. Beside him on another gaudy chair sits the Bishop, as last of the Barons of the Island, and the Deemsters and Council stand beside them. On the second platform stand the twenty-four members of the House of Keys (the world's smallest Parliament), on the third the clergy, High Bailiffs and members of the Bar, and on the fourth or lowest platform the minor officials. The First Deemster then reads the English titles of the Acts which have been passed and received the Royal Assent during the year, which are then read in Manx, and the Acts of Tynwald are the law of the land.

Then the Coroners, on bended knees, deliver to the Governor their wands of office and receive them back again, and the grandees march back to the Church. Mr. William Radcliffe, in his *Guide to the Isle of Man*, quotes an interesting old document on the Manx Constitution, which I append :

'This is the Constitution of old time, how yee shall be governd on Tynwald Day. First you shall come hither in your array, as a king ought to do, by the perogatives and royalties of the land of Manne. And upon the Hill of Tynwald sitt in a chaire covered with a royall cloath and cushons, and your visage unto the east, and your sword before you holden with the point upwards. Your Barrons in the third degree sitting beside you, and your beneficed men and your Deemsters before you sitting, and your clarke, your knights, esquires and yeomen about you in the third degree, and your worthiest men in your land to be called in before your Deemsters if you will ask anything of them, and to hear the government of your land and your will; and the commons to stand without the circle of the Hill with their clarkes in their surplices. . . .

'Then the chief Coroner, that is the Coroner of Glenfaba, shall make affence upon pain of life and lyme, that noe man shall make any disturbance or stirr in the time of Tynwald, or any murmur or rising in the King's presence, upon pain of hanging or drawing, and then to proceed in your matters that touch the government in your land of Manne.'

This is dated somewhere about 1417, and with modifications due to altered conditions this procedure is still followed, but only brief extracts of the Acts are now read in Manx and English.

July 12th

THE VINTNERS' COMPANY:
ITS WHITE SMOCK PROCESSION AND ITS SWANS

A VERY picturesque old custom is still observed in the City of London on July 12th by the Master, Wardens and Members of the Worshipful Company of Vintners. They walk in solemn procession from their hall in Upper Thames Street to attend service at the Church of St. James, Garlickhythe.

The procession is led by two wine-porters in old-time top hats and white smocks. They sweep the streets with besom brooms, and with the same implements order taxis and lorries out of the way of the procession. Then comes the beadle, a magnificent figure in black and gold livery, followed by the stavesman, swan-marker and bargemaster. Behind these officials follow the members of the Company of Vintners, some in morning dress, and others in brown and mauve robes. As they walk they sniff from time to time posies of sweet herbs and flowers. This, of course, was originally done, because of the garbage in the uncleaned streets, which the broom-men in front of the procession could not possibly remove. The swan-marker reminds us that the Vintners still share the right to own swans on the River Thames, which in former times was a very great privilege, as the swan was a royal bird. Even today all swans on the river are regarded as Crown property except those belonging to the Vintners' Company and the Dyers' Company. Each of these maintains a Swan Warden, and these two, in conjunction with a third Warden representing the Crown, perform their 'Swan Upping' duties on the last Monday in July. The unfortunate birds are rounded up and marked by nicks on their bills, to determine their ownership. It is good fun to watch, as a swan is a lusty bird and can be very troublesome, while a blow with its wing has been known to break a man's leg. Add to this a snaky neck with a horribly sharp bill at the end, stabbing in all directions and hissing like the Scotch Express, and you will agree that on that Monday, at any rate, the Swan Uppers earn their pay. The nicks are cut in the bill with a sharp knife, and the number of nicks determines the ownership. Hence the well-known inn sign 'The Swan with Two Necks': this should be 'Two *nicks*.'

The City Companies are very tenacious of their ancient customs and privileges. There are a number of these companies – the Fishmongers', the Merchant Taylors', the Skinners', the Goldsmiths', the Gunmakers', the Apothecaries', the Stationers', the Weavers' and the Girdlers' being among the most important. The last-named guild retains the custom of crowning their Master with a remarkable head-dress somewhat resembling a royal crown, while the Wardens also wear handsome head-dresses of a different shape.

AUGUST

PUNCH AND JUDY

*WARDMOTE OF THE WOODMEN OF
  ARDEN, MERIDEN, WARWICKSHIRE*

THE LORE OF HARVEST

RUSH-BEARING AT GRASMERE

HIGHLAND GAMES AND GATHERINGS

PUNCH AND JUDY

IT is during the month of August that we can usually find our old friend Mr. Punch, busily entertaining a crowd of children on the sands. Though its popularity has greatly diminished in these modern days of cinemas and wireless, the show still delights the children, and is of interest to those ' of riper years.' For here we have a survival of the ancient puppet shows and possibly traces of the pre-Reformation mystery and miracle plays.

Figures with movable limbs have been found in the tombs of Egypt and Etruria ; and plays worked with moving figures are of very ancient origin and still survive in Egypt, India and the East, where they are known as ' Ombres Chinois ' or ' Shadow Plays,' and are usually coarse and revolting to European tastes. The general style of such figures is usually a loose dress permitting a man to put his hand inside, the arms are worked by the thumb and second finger, and the head by the index-finger – though some are worked instead by wires. So far as the *name* Punch is concerned, it undoubtedly comes from Italy, Punchinello being a Neapolitan clown. F. Galiani tells us that Puccio d'Aniello was a wine-grower of Acerra, near Naples, who by his quick wit and grotesque appearance vanquished a troupe of strolling comedians in their own sphere, and was induced to join the company. After his death his place was taken by a masked actor who imitated his dress and manner. There are various other theories as to the origin of the character; those who think the show derives from the mystery plays suggest that Punch is Pontius Pilate, and Judy presumably the traitor Judas[1]; and there can be no question that Punch bears a strong resemblance both to the Vice of the ancient Moralities, and to the classical Maccus. In 1727 a bronze statue of the latter was discovered, and the hump back and large nose so characteristic of Mr. Punch are very obvious. This reminds us of the fact that Mr. Punch is a very amorous gentleman, and there is a widespread popular idea that a large nose indicates another personal defect which I will not specify, but which the Italians have crystallised into a crude proverb.

While it is probable that puppet shows with sacred characters existed in England before the Reformation, and certainly continued through the reign of Queen Elizabeth, we do not find the actual name Punch earlier than 1703, when he appeared in a puppet play of the Creation of the World at St. Bartholomew's Fair. In his book on *Punch and Judy* Mr. George Cruikshank tells us that Fleet Street and Holborn Bridge were much used by puppet shows during Elizabeth's reign, and Fleet Street was 'infested' by 'motions' and 'monsters' right down to the

---

[1] A difficulty about this theory is that Judy was called Joan at first. My own view is that the names Punch and Judy have *nothing whatever* to do with the mediæval ethical drama.

Restoration. It reached the zenith of its popularity in the eighteenth century, when there was a letter in the *Spectator* supposed to have been written by the sexton of St. Paul's Church, Covent Garden, complaining that the performance of Punch and Judy thinned the congregation of that church! It was even asserted that it entered into serious competition with the Opera, and reduced the receipts of that august institution! I will close by quoting one or two verses from a ballad called 'Punch's Pranks,' written at the end of the eighteenth century.

> 'Oh, harken now to me awhile,
> A story I will tell you
> Of Mr. Punch, who was a vile
> Deceitful murderous fellow:
> Who had a wife, a child also –
> And both of matchless beauty;
> The infant's name I do not know,
> Its mother's name was Judy.
>   *Right tol de rol lol, etc.*
>
> But not so handsome Mr. Punch,
> Who had a monstrous nose, sir;
> And on his back there grew a hunch
> That to his head arose, sir.
> But then, they say that he could speak
> As winning as a mermaid;
> And by his voice – a treble squeak –
> He Judy won – that fair maid.
> But he was cruel as a Turk;
> Like Turk was discontented,
> To have one wife, t'was poorish work –
> But still the law prevented

His having two, or twenty-two,
Though he for all was ready;
So what did he in that case do?
Oh, sad – he kept a lady.'

The ballad is interesting as it sets out the 'plot,' if such we may term it, of the Punch drama as it was played in my childhood. When Judy discovered the kept lady, she was very jealous and pulled her husband's nose and also assaulted the erring fair. He thereupon killed her with his bludgeon, and threw his child out of the window. When his wife's relations expostulated with him he knocked them down too. After a career of amorous adventure he was finally clapped into prison and sentenced to be hanged. On the scaffold, he pretends that he does not know how to put his head into the noose, and the hangman kindly shows him the proper way. Mr. Punch promptly pulls the rope and hangs the executioner. This usually finishes the entertainment, but in former times there was a last scene in which the Devil comes after the victorious old reprobate, and is himself killed by Mr. Punch. The ballad closes with these words :

'The Devil with his pitchfork fought,
While Punch had but a stick, sir,
He kill'd the Devil as he ought.
Huzza! There's no Old Nick, sir!'

While I have the utmost respect for the learned professors who inform us that this 'drama' is derived from the mediæval religious plays, I am bound to say that the passage of the years has effectively eliminated all religion and morals and left us a most disreputable story!

We get a much better idea of the ethical folk-drama from the Mummers' Play which I shall fully describe on pages 217–233.

Month of August

WARDMOTE OF THE WOODMEN OF ARDEN, MERIDEN, WARWICKSHIRE

MERIDEN is a pretty, straggling village, well known to motorists on the Holyhead road. It claims to be the very centre of England, and it is a very interesting fact that water flows from a pond on high ground in the parish in two directions, and so reaches the Bristol Channel to the S.W. and the Humber to the N.E. The village is almost in the centre of the once mighty Forest of Arden, which covered the whole of Warwickshire and parts of adjacent counties.

' That mighty Arden held ever in height of pride,
  Here one hand touching Trent, the other Severn's tide.'

Owing to the extent and importance of this great forest, the Woodmen were formerly an order of some consequence, and for centuries held their annual meeting at Meriden. This is still celebrated every year in August, and includes a great archery contest. There are picturesque costumes, and quaint ceremonial, and the whole affair is a most interesting survival.

THE LORE OF HARVEST

THE invention of the mechanical harvester has greatly simplified the task of the farmer and has advanced the Harvest Home by nearly a month. In a fine summer the harvest is sometimes finished before the end of August in southern England, whereas in my own boyhood it was seldom over before the middle or end of September. Even this was an advance, as we have it on record that the Druids held their 'Harvest Festival' on November 1. From time immemorial the safe ingathering of harvest has been regarded as an occasion for rejoicing, though, of course, the actual date varies widely. The ancient Jews, for instance, held their harvest rejoicings at the Feast of Pentecost.

Even as recently as a generation ago, the festival of Harvest Home was everywhere celebrated by a supper for the farm-workers, with much merriment. In many districts it was also an occasion for rustic love-making; if a youth 'named' a maiden for a toast, this was taken as a declaration of his interest in her, and if she accepted the compliment they would forthwith commence 'walking out.'

That brilliant, extraordinary and most unfortunate man, Eugene Aram, devoted considerable study to the question of the 'Mell Supper,' as the Harvest Feast was called in his day. His opinions are supported by our later knowledge, and suggest that these usages are relics of pagan ceremonies, and are of far greater antiquity than had been previously supposed. What an amazing thing that so gifted and cultivated an individual should have been guilty of the murder of a shoemaker for a comparatively small sum of money ! Still more remarkable were the circumstances which followed the crime. No hue and cry was raised, and Aram was not even suspected of having had any hand in Clarke's disappearance. Meanwhile he obtained a post as tutor, and became a profound student of the dead languages, learning Greek, Hebrew, Celtic, Arabic and Chaldee. Then, at the zenith of his career, the bones of his victim were accidentally discovered, and he found himself in the condemned cell.

But I must return to our Harvest Lore. Until comparatively recent times there were considerable vestiges of primæval usages in connection with the Harvest. In Northumberland, when the last sheaf was set up, the reapers were wont to shout, " We have got the kern," and the sheaf was dressed in a white frock, decked with coloured ribbons, and hoisted on a pole. This was known as the ' Kern Baby ' or Harvest Queen, and was carried to the Harvest Festival and set up in a prominent place. Kern Baby is of course ' Corn Baby,' and is a trace of the old animistic belief in the Corn Spirit or Corn Mother.

In Scotland the last sheaf cut before Hallowmas is called the ' Maiden,' and the youngest girl in the field cuts it ; but if harvest does not finish

until after Hallowmas the last sheaf is called the
'Cailleach' (or old woman). In some cases this
last sheaf was supposed to have specially nourishing properties; in others it was carefully stored
in the farmhouse till the next harvest came round.

In Scotland the celebration of the 'Klyack'
is still kept up, particularly in Aberdeenshire.
The 'Klyack' is, of course, another name for
the last sheaf, and in order that a worthy specimen may be available, the largest and heaviest
sheaf is usually laid aside. When the harvest is
safely gathered in, the workers return to the field
and bear the 'Klyack' in triumph to the farmhouse. There it is bedecked with ribbons and
hung up in a conspicuous place to await subsequent rejoicings.

When the first of the season's grain has been
ground by the local miller and returned to the
farm, a quantity of this new meal is placed in a
tub and ale is poured over it. A couple of bottles
of whiskey are added to the mixture, and the
neighbours are summoned to the feast, which is
known as 'Meal and Ale.' Every farm has its
celebration, and neighbours arrange together to
prevent two such occurring on the same night.
When the guests arrive, the tub of meal is placed
in the middle of the kitchen floor, and is freely
partaken of; when it is finished the party goes
out to the barn and dance beneath the 'Klyack'
until the early hours of the morning. This reminds us of the Roman harvest, presided over
by the goddess Ceres. The 'Klyack' itself is

stored safely away until the end of the year, when it is threshed and given to the birds, which is supposed to bring good luck in the coming year.

It is much to be regretted that these picturesque old customs are dying out, as the Harvest Home was not only a delightful survival in itself, but fostered that good feeling between master and men which is certainly more necessary to-day than ever before.

RUSH-BEARING AT GRASMERE

THE traditional day for this interesting and beautiful old custom is the Feast of St. Oswald on August 5th, but in order to meet the convenience of the crowds of week-end visitors who desire to be present, it is now celebrated on August 5th, if a Saturday, or otherwise the nearest Saturday. It is hard to conceive of a lovelier setting for a picturesque custom. The old, square-towered church – where Wordsworth worshipped, and he and his family are buried – stands close beside the lake, cradled in glorious mountains. Though almost the smallest, I consider Grasmere the loveliest of the lakes, and happy is he who shall visit it in bluebell time.

Rush-bearing is a survival of the old custom of strewing rushes on the floor before the days of carpets. We have a record that Thomas à Becket was regarded as fussy in his time, because he insisted that the floor rushes should be changed daily, and from what we know of domestic habits in his period, I consider the saint was a wise man. Though there is no longer need for them as floor covering, the rushes are carried to Grasmere Church for the festival, which in many ways resembles the May Day processions. The children are much in evidence at this festival of youth; many are themselves adorned with garlands of flowers, and they carry between them linen sheets beautifully decked with flowers and filled with rushes. There are also ornaments made from

plaited rushes, sacred symbols, crosses, triangles
and harps, and even more elaborate works of
art, such as Moses in a basket woven from rushes.
This beautiful custom has fostered a local craft
of rush-weaving, which reminds us of the still
more remarkable art of flower-petal pictures at
Tissington (see pages 80–84). The interior of the
church is a wonderful sight. The floor is strewn
with sweet-scented rushes, and the building
itself is beautifully decorated with floral designs.
The ancient rush-bearers' hymn might be quoted :

> ' Our fathers to the House of God,
>   As yet a building rude,
>   Bore offerings from the flow'ry sod
>   And fragrant rushes strew'd.
>
> There of the Great Redeemer's grace
> Bright emblems here are seen !
> He makes to smile the desert place
> With flowers and rushes green.'

There is a somewhat similar observance at
Killin at Christmas (see page 214). The celebrated Grasmere Sports, with a wonderful display
of wrestling, take place on the third Thursday in
August.

The secluded vale of Grasmere is one of the
few places where the old Westmorland dialect
still exists, though Council Schools are rapidly
killing it, and here in January every year there is
performed a remarkable folk-play. Players are
locals who are largely selected for their ability to
speak the vernacular, and the central character in

the play is usually a Statesman. These sturdy
farmers, toiling on their own land, are a splendid
race for which Lakeland has long been famous,
and it would be a fine thing for British agriculture
if there were more owner-occupiers of this type.
The action of the plays – for there are several to
choose from – is placed in the bustle of the farm
kitchen, or the more sedate farm parlour. Last
year the play was *Lad's Love*, and included the
singing of those famous Lakeland songs 'John
Peel' and 'Sally Grey.'

August – September

HIGHLAND GAMES AND GATHERINGS

THOUGH scarcely folklore, the celebrated Highland Gatherings, with their display of games, Tossing the Caber, Highland Dances, etc., certainly demand mention here. They are not only of the deepest interest, but are held in a setting of the most splendid scenes of mountain grandeur that can be found in the North. The premier meeting is that of Braemar, usually held in August in the Princess Royal Park, and generally attended by a royal party from Balmoral. Space does not permit me to describe these interesting functions, but I will briefly give the dates for the present year (1930) and the places where they are held.

Aug. 2 : Strathallan Highland Games, Bridge of Allan.
 ,, 29–30 : Cowal Highland Gathering, Dunoon.
 ,, 26 : Lochaber Games, Fort William.
Sept. 18 : Northern Meeting, Highland Games, Inverness.
 ,, Second week : Braemar Gathering.
    Argyllshire Highland Gathering.
    Aboyne Games.
 ,, Third week : Highland Gathering, Blair Atholl.

SEPTEMBER

THE DANCE OF THE DEERMEN
SOME OLD ENGLISH COUNTRY FAIRS

First Monday after September 4th

THE DANCE OF THE DEERMEN, ABBOT'S BROMLEY, STAFFS.

THE little town of Abbot's Bromley, about six miles from Rugeley (Staffordshire), is widely famed for its extremely interesting survival, the Horn Dance, which takes place in the Market Place of the village. The dance is very ancient, some writers say it is four hundred years old, but I am inclined to think it is much older, for two reasons : firstly, because of its close resemblance to the Morris Dances and to primitive customs found widely among savages who hunt animals ; and secondly, because of the remarkable circumstance that the horns are *reindeer* horns, and were the same when first described by Plot (*circa* 1680). Now reindeer became extinct in Britain before the arrival of the Normans, so the dance must have been observed in Saxon – or still earlier – times, or the horns must have been imported from abroad. The idea of a folk usage based on an imported article is absurd, since it naturally gathers around something native and familiar. I will quote briefly from Plot's description of the dance as he saw it nearly four hundred years ago :

' A person carried the image of a horse between his legs, made of thin boards, and in his hand a bow and arrow, which passing through a hole in the bow and stopping upon a shoulder it had in it, he made a snapping noise as he drew it to and fro keeping time with the music. With this man danced six others, carrying on their shoulders as many reindeer's heads, three painted white and three red, with the arms of the chief families (Paget, Bagot and Wells) depicted on the palms of them, with which they danced the hays and other country dances. To this Hobby Horse Dance, there also belonged a pot, which was kept in turns by four or five of the chief of the town, whom they called Reeves, who provided cakes and ale to put in this pot. All people who had any kindness for the good intent of the sport giving pence apiece for themselves and their families, and so foreigners too that came to see it ; with which money (the charge for the cakes and ale being defrayed) they not only repaired their church, but kept their poor too.'

The dance is still performed annually and much as described. The performers are now twelve in number, of whom six have the deer horns, and the remainder are made up as follows. A man called Robin Hood with the Hobby Horse, a man dressed as Maid Marian, one with cross-bow and arrow, a fool and two musicians. All wear quaint costumes of red and green and flowered breeches, and there is much snapping of the hobby horse's wooden jaw, and bow and arrow. The relics are kept in the church tower when not in use.

Mr. Charles Masefield suggests that the origin of the dance was to prove that the townspeople possessed the right of hunting deer in the adjoining Needwood Forest. This is interesting and plausible, but I am of opinion that the survival is of very much earlier date.

SOME OLD ENGLISH COUNTRY FAIRS

DATES : September and October are the best months for the old fairs, many of which are still kept up. The following may be noted.

SEPTEMBER : Woodbury Hill, near Bere Regis (Dorset), Tiverton (Devon), Barnet Horse Fair, Widecombe Fair (Devon).

OCTOBER :
Oct. 1 : Mop Fair, Stratford-on-Avon.
Oct. 3 : Goose Fair, Nottingham.
First week in Oct. : Weyhill Fair, near Andover, Hants.

THE old English Country Fair is one of the most delightful and characteristic relics of the past which still remains to-day, though, alas, it is rapidly passing away and may disappear entirely in the next generation. The Fair is one of the most ancient institutions in this country ; it may have originated during the Roman occupation and there are considerable traces of it in Saxon times.

At first religion and pleasure were the only elements in these festivals, but as great numbers of people attended, hawkers and pedlars came to sell their petty wares ; and later the merchants also, who set up stalls and booths in the churchyards. It is probable that the festivals of the Saxon gods were adapted to a new form, Christian in name, but heathen and grossly sensual in practice, traces of which lingered on till a century or so ago in the form of the country wake or revel. The Venerable Bede quotes a permission granted

by Pope Gregory to St. Augustine to make use of
the Saxon temples as Christian churches, removing
only the idols. He also says that on the
day of the dedication, or at the nativities of the
holy martyrs, cattle may be slain for feasting, and
huts may be erected round the former temples
for the accommodation of the people who resorted
thither. (Bede : *Ecclesiastical History*, Bk. 1.)

In these feasts we find the germ of the great
country Fair, and the greater the reputation of
the saint, the larger the multitudes who resorted
thither. The morning would be spent in religious
services, and the remainder of the day in dancing,
wrestling, cudgelling, drinking and crude love-making.

The holding of these festivals (which by the
addition of the trade element had really become
Fairs) on Sundays was objected to by the clergy ;
and in the reign of King John the Abbot of Ely
attempted a prohibition without success, though
it was finally carried by Henry VI. A Statute of
Edward I commanded that neither Fairs nor
Markets should be kept *in churchyards*.

Meanwhile the Fairs had become enormously
valuable, and the Church was quick to secure
a new source of unearned increment. Markets
were a prerogative of the Lords of Manors,
and being held weekly were a valuable property.
Fairs, however, belonged to the Crown, and
clerical influence at Court speedily obtained
numerous charters to hold Fairs and levy tolls on
the dedication feasts of churches and monasteries,

or on popular saints' days. Some celebrated examples of the latter include St. Bartholomew's Fair at Smithfield, St. George's Fair at Modbury, and St. Giles' Fair at Winchester and Oxford. Of the last two, the former has now disappeared, and the latter has degenerated into a noisy pleasure fair, but in the fourteenth century St. Giles' Fair at Winchester was only rivalled in Europe by that of Beaucaire in Languedoc. Vast numbers of merchants came from abroad, as well as hordes of pilgrims *en route* for the Shrine of St. Thomas at Canterbury. It should be mentioned that both this Fair and that at Shalford (on the Pilgrim's Way in Surrey) were dated to catch the pilgrims on their journey.

All the vast revenues of St. Giles' Fair, Winchester, went into the pockets of the Bishop, and it is interesting to find that for the sixteen days that it lasted all trade was prohibited for seven leagues around. So when both Winchester and Southampton were crowded with travellers, the shopkeepers were obliged to close down, and all the profit went to the Bishop. Nor was this all. For the period of the Fair, the Court of Pie Poudré superseded the regular courts and exercised drastic sway over the Fair and Fair-goers. It is probable that even the local innkeepers derived little benefit from the hordes of visitors to St. Giles' Fair, since both food and lodging were provided at booths in the fair itself, and ladies of easy virtue were numerous. Some of the Fairs became celebrated for the fare provided, and a staple meat gave its name to the Fair itself, e.g. Goose Fairs at South Brent, Tavistock and Nottingham; Lamb-pie Fair at Buckfastleigh in Midsummer, and Pear-pie Fair at the same place in September; Totnes Gooseberry-pie Fair in May. The term Mop Fair implies that it is a hiring fair, i.e. one to which men and women resorted to obtain employment. This is still done at some old fairs and markets, and a visitor to Salisbury Market in August or September can see a rosy-cheeked farm labourer standing about with a whip-lash or bit of sheep's wool pinned to the lapel of his coat. He is looking for a job, and the whip-cord signifies that he is a carter, or the wool a shepherd.

Other fairs became celebrated for the merchandise sold there, such as Bristol Wool Fair, Cheddar Cheese Fair, Exeter Leather Fair, Weyhill Sheep and Hop Fair, Overton Sheep Fair, Barnet Horse Fair, Bampton Pony Fair etc; most of these are still carried on, and are full of interest.

For centuries the ordinary licensing laws were abrogated during the period of the annual fair, and a bunch of evergreen hanging in front of a house permitted the owner to sell liquor. Though this privilege has now vanished, considerable extensions of time for the consumption of alcoholic refreshments are still permitted by the magistrates in market hours. At Tiverton the opening of the Fair is attended by the Mayor and Aldermen in their robes, accompanied by the

Beadle with his Mace and the Town Clerk in his robes of office. Pennies are scrambled to the children in the main street, and the Fair is formerly declared open, though to-day it is a mere shadow of its former self. There is a chance, however, that the Fair may become more valuable in future, so every effort is made by the Civic Fathers to retain their ancient charter, and the Opening Ceremony is punctiliously performed every year on October 3rd. A similar ceremony is annually performed at Ely, when 'all vagabonds, idle and misbehaving persons, all cheaters, cozeners, rogues, sturdy beggars and shifters' are commanded in the name of the Bishop to depart immediately out of the Fair. At Newcastle the Mayor and Sheriff proclaim the opening of the Fair, and at Modbury (S. Devon) the Portreeve performs this office. There is a very ancient Fair at Corby (Lincs.) held on the Monday before October 11th, which at one time was the largest sheep fair in the country; and at Corby (Northants), eight miles from Kettering, there is a most remarkable Charter Fair, held every twenty years (1902–22–42). It dates from a Charter of Queen Elizabeth given in the twenty-seventh year of her reign, and later confirmed by Charles II. To-day it has become a modified saturnalia, but incorporates some interesting survivals. Every entrance to the village is blocked during the Fair by closed gates, over which is the legend 'God Save the King,' and a toll is demanded of every person who seeks

admission. Those who refuse to pay are confined
in the stocks (this is the only survival of their use
in England). Women refusing to pay the toll are
carried to the stocks in a chair, but male defaulters are set astride a pole carried by two
strong men, which is a precious opportunity for
some rough horseplay. The Charter, which is
solemnly proclaimed at the opening of the Fair,
gives the power of imprisonment for a certain
brief period if the toll is not paid.

There is another very ancient Fair on May 8th
at Helston (Cornwall) which may date back to
Roman times. The celebrated Furry Dance (see
pages 76–9) is performed in connection with this
annual event.

I have pointed out that the Fairs were first
religious, then recreative, and finally commercial.
The first disappeared at the Reformation ; and
the last received a mortal injury from the invention of the railway. This rendered them largely
uneconomic, since it was usually more profitable
to convey produce direct to the large cities where
it would be consumed, than to cart it to a fair.
But though the blow was mortal, it has not
proved immediately fatal. As recently as the
'eighties west country farmers were still carrying
cheeses to Weyhill Fair in order to exchange
them for hops from East Hants, Sussex or Kent ;
and even to-day some business is done at the
Fairs. The Sheep Fairs are well supported, and
will no doubt continue ; the Horse Fairs, though
sadly diminished, still survive. Many quaint old

craftsmen and sellers of old-world things are even now to be seen there. One old gentleman I have often met makes and carries round those ancient, long, churchwarden pipes of baked clay. Another, till recently, made and sold the wooden shovels beloved of maltsters, but malt, alas, is disappearing before the advent of chemical beer brewed from sulphuric acid! Old foods and drinks can be discovered, such as the furmity mentioned by Thomas Hardy in *The Mayor of Casterbridge*, and queer gingerbread 'fairings' and brandy snaps can still be purchased, as well as the old-fashioned lollipops which delighted our grandfathers but raise little enthusiasm now. A few of the old touring Theatrical Companies still survive, where *The Tragedy of the Red Barn* or similar horrors are enacted, but the competition of the cinema is rapidly killing the Pleasure Fair. Even the strident tones of the steam organ and the lure of the 'roundabouts' fail to attract as of yore. A generation ago I have seen a farmer's young son spend half a sovereign on two or three girls at a country fair; but to-day the Morris car carries him swiftly to the more sophisticated delights of the market town. There was money in the 'roundabouts' once. I well remember a tattered old gipsy, who lived in rags and died on a wretched pallet in a leaky tent. When the filthy mattress was examined after his death, £15,000 in gold was found hidden in it; all of which he had earned in visiting country fairs with his roundabouts and other delights.

But though the old country Fair has almost lost its trade, international Fairs are by no means dead. This is proved by the success of the great British Industries Fairs at London and Birmingham and the enormous German Fair at Leipsic; but all of these are Fairs for wholesale, not retail, buyers. The world's greatest Fair up to the outbreak of war was that of Nijni-Novgorod, which resembled on a much greater scale the well-known Sturbridge Fair near Cambridge in mediæval days. The Nijni Fair dates from the fourteenth century, and in the 'eighties of the last century had grown to such an extent that silks and cotton goods alone were sold to the tune of over four millions sterling. In 1697 the whole trade was £12,000, and a century later had grown to four and a half millions, and to twenty millions by 1874; six thousand shops were let in this Fair, the stalls, shops, booths etc., covering about eight square miles; while there were about ten miles of wharves on the river frontage.

But I must return to our English Country Fairs. To the student of human nature their greatest charm is the people. All rural society is there: the shepherd and the carter, the tradesman and the squire, the cheap-jack, the pill-doctor, the corn-cure man; and last but not least the Romany Folk. What a romance attaches to the name of this picturesque, idle, decaying race!

'A vagrant crew, far straggled through the glade,
With trifles busied, or in slumber laid.
Their children lolling round them on the grass,
Or pestering with their sports the patient ass.
The wrinkled beldame there you may espy,
And ripe young maiden with the glossy eye.
Men in their prime, and striplings dark and dun,
Scathed by the storm and freckled with the sun;
Their swarthy hue, and mantle's flowing fold,
Bespeak the remnant of a race of old.
Strange are their annals – list and mark them well –
For thou hast much to hear, and I to tell.'

No fair or country fête is complete without the Romany Folk, with their gaily painted caravans, and rude tents exactly like those I have seen used by wandering tribes in Africa. Very often they are the proprietors of the swings, roundabouts and other delights of the fair, and if horses are for sale we shall see dark-hued gipsy lads showing off their paces. For horse-dealing and horse-faking too none can equal the Romany Folk; and the gipsy wife has a well-deserved fame as a fortune-teller. I recently had a long conversation with Mary Lee, one of the famous family of Lees. It was another member of this family who so correctly told Queen Victoria's fortune at Epsom. She speaks the old Romany tongue fluently, far too fluently for me, I need hardly say. I approached her with diffidence and very timidly said, "Howie been baishen? Can you rakker Romani?" ("How do you do? Can you speak Romany?") The result was a torrent of that

ancient tongue that fairly left me gasping for
breath. I only know a few words of the language,
but even a little knowledge helps one to distinguish between the Romani Chal (or true
gipsies) and the despised Churdi (or half-breeds)
who also roam the country. The genuine gipsy
families are very proud of their lineage, a few of
the famous names being Lee, Cooper, Lovell,
Stanley, Boswell, Hearne, Chilcott, Young and
Worton.[1] The Dennards and Brazils are two
famous families of wicker-workers. It is not
generally known that there are six camping
places in the New Forest where the gipsies are
free to reside, and there are also several such in
the Forest of Dean. At the time of writing (1930)
there is a gipsy King living in the Forest of Dean.
His name is William Johns and he will be 90
next birthday. His Queen (who had been his
faithful companion for over sixty years) died in
1929. The old gentleman was recently visited
by a Press representative, to whom he sadly
remarked, " Our race is fast dying out ; the rising
generation in these days does not take to the
roads. In a few years' time there will be scarcely
any of our encampments left." Gipsy ' Kings'
are a romantic subject, though coronations are
seldom heard of nowadays. One of the last
instances occurred towards the middle of the last
century at Yetholm. I quote the following extracts from the *Kelso Chronicle*.

[1] A few more gipsy names are : Baker, Baillie, Barrington, Blewitt, Broadway, Buckland, Buckley, Burnett, Carew, Carter, Faa, Fletcher, Glover, Jowles, Jones, Smith, Stokes, Stevens.

MARY LEE OF THE NEW FOREST.

To face p. 182.

'The coronation of her majesty Queen Esther Faa Blyth ... took place last week. ... Esther was accompanied by princes and princesses of the royal blood – her brother Prince Charles, and nephew of the same name and title; and two of the princesses attended Her Majesty on horseback, some of her majesty's grandchildren also being present. The queen, mounted on her palfrey, proceeded to the cross, the crown-bearer and crowner following behind. The procession having halted, the crowner stepped forward and placed the coronet upon her head, a Scotch thistle being a prominent object on it. The crowner, from a roll of parchment, proclaimed that he, having proclaimed her deceased father King Charles, from his inherent right as crowner and from the fact of the late King dying intestate, now placed the crown on the head of Esther, and with public challenge at the cross of her dominions, he proclaimed her Queen Esther Faa Blyth, "Challenge who dare" (loud cheers).'

The genuine gipsy Kings and Queens are selected from certain ancient or royal families. I have discovered a number of their graves. At Highclere (Hants) there is a tomb bearing these words:

> 'Sacred to the memory of RICHARD STANLEY, who died December 15th, 1822, aged 21 years.
>
> 'The widowed mother and the orphan dear
> Their heavy loss together now deplore.'
>
> 'KING OF THE GIPSIES.'

At Calne, Wilts, there is a stone built on to the wall of the Parish Church bearing the romantic words 'King of the Gipsies.' Though the inscription is somewhat defaced, I believe I am correct in stating that the grave is that of Inverto Boswell, who died in February 1774. It is said that he was interred with mystic heathenish rites by the tribe.

Other gipsy Kings may be briefly referred to thus:

In the island of Orkney, William Nowland, who died March 1884;

St. Anne's Churchyard, Lewes, Charles Lee, who died August 1832;

At Beckenham, Margaret Finch, who died in 1740;

At Ickleford, near Hitchin, Herts, Henry Boswell;

At Beaulieu, Hants, Joseph Lee, who died in 1844;

At Launton, Oxon., James Smith, who died in 1830.

It is probable that few tombs will now be discoverable, but at the time of writing the inscriptions of the Highclere and Calne stones are clearly legible. There is another interesting tomb at Yatton (Somerset) to Merrily Jowles, the inscription on which was written by her husband, a scissor-grinder :

> 'Here lies Merrily Jowles,
> A beauty bright,
> She left Isaac Jowles,
> Her heart's delight.'

These touching words remind us of the fame of gipsy maidens for their beauty; and though much exaggerated by writers of fiction, it is a fact that gipsy maidens are often beautiful. In France at the great annual Gipsy Festival the most beautiful girl is elected Queen for the period of the festival, and I was fortunate enough (in May 1928) at this festival to see and photograph the really lovely young girl who was thus honoured by the tribe. She was being sketched by a famous artist, when I came up, and I was able to secure a photograph of Her Majesty with her sister. She was slim but shapely, and had glossy dark hair and a lovely pink and white complexion, which the open-air life will ruin in a few years. I have also seen and photographed some comely wenches among the celebrated *gitanas* of Granada; but most of these to-day are ill-favoured harridans, with the morals of a mediæval camp-follower and the features of a weather-beaten trollop. These harpies made great efforts to lure me inside their houses and promised strange delights; but I kept my money in my pockets and stayed in the street. Then they swarmed round me and demanded ten pesetas merely to pose for a photograph! The gipsies of Granada have a bad reputation throughout the Peninsula, and in religion and morals remind us of the Wallachian view of the race as exemplified by their proverb, 'The gipsies' Church was built of bacon, and the dogs ate it up!' A Hindu proverb says, 'There are seventy-four and a half religions in the world, and the half belongs to the gipsies.' It is certainly a fact that, wherever they are found, the gipsies seldom take much interest in religious observances, and if they conform at all it is almost always to the religion of the country of their adoption.

One romantic tradition of their origin states that they are descended from Samer. He was the goldsmith who made the Golden Calf which the Israelites worshipped in the desert with such tragic results, and as a punishment he and his accomplices were turned out of the tribe and have been wanderers ever since. This legend explains the name Gipsy, i.e. Egyptian. The Romany folk hate the term, and usually call themselves ' Travellers.' It was long thought that the gipsies really were of Egyptian origin, and perhaps were descended from the people who followed the Children of Israel out of Egypt (Exodus xii. 38) – ' A mixed multitude went up also with them.' But it is now generally accepted that they came from India. Grellman has pointed out that twelve out of every thirty words in their language are either pure Hindustani or bear a striking resemblance to it ; and there is little doubt that they were originally among the Sudras and Pariahs of Hindustan. This is confirmed by their language, their lack of religious belief, their fondness for wandering, their love of horses, and their skill as tinkers ; in all of which they resemble the Sudras. Their swarthy complexions and love of flashy colours and showy beads, etc., are also oriental; and I have seen some gipsy girls who resemble the Ouled Naïls of the Sahara both in appearance and dress.

It is probable that the Sudras, who were the progenitors of the gipsies, fled from Hindustan to escape the cruelties of Timur Beg. He was a tyrant who ravaged India with fire and sword and killed half a million of the natives in the attempt to convert them to his own religion. They flocked into Asiatic Turkey and passed into Europe via the Balkans, while others seem to have crossed the Isthmus of Suez and travelled into Egypt, where they lived a nomadic life for some time. Prior to entering Europe they divided themselves into small bands and were reported in many cities during the fifteenth century. They spread all over Europe, and excited much interest and considerable opposition by their uncouth appearance. They must have been a rough crowd. The women and children, clad in tattered finery, travelled in rough carts, and the swarthy unkempt men trudged alongside; while swarms of half-clad older children ran ahead and attracted the attention of passers-by to their feats of jugglery, craft and deception, by shrill cries and monkey-like grimaces. They were expelled from some countries and hunted down like wild beasts in others, but still they went on. Sometimes they gained an entrance by cunning, as at Bologna in July 1422, when one of them, styling himself Duke Andrew, headed a band of a hundred followers and presented an alleged decree from the King of Hungary permitting him to rob without penalty for seven years, as a reward for re-baptism into the Christian faith. Others made a pilgrimage to Rome and received in return a passport from the Pope, calling on the faithful to give them alms. Another cunning rascal, one Anthony Gawine, styled himself Earl of Little Egypt, and not only visited Scotland with his band, but obtained from King James IV a letter of introduction to the King of Denmark when the artful Anthony had tired of the thrifty Aberdonians. King James warmly recommended them to his royal Uncle of Denmark, stating that as they were wandering Egyptians, they must be better known to his Danish Majesty than himself, since Egypt was nearer to him! Though James IV assisted the gipsies, James V fell foul of them. It happened thus. The king, when travelling in disguise, fell in with a band of the Romany Folk near Wemyss and joined in their carousals. On attempting, however, to take liberties with one of their women, he was struck on the head with a bottle and otherwise roughly handled. He escaped from them with difficulty, and immediately issued an Order in Council authorising the execution of any gipsy, simply for being a gipsy. This barbarous treatment continued for some years in Scotland, and to a less extent in England. King Henry VIII ordered them to leave the kingdom, and Mary ordered that any of them remaining in England for more than one month, or even any non-gipsies consorting with them, should be put to death. Even as late as the seventeenth century this brutal persecution continued, and Judge Hale tells us that at one Suffolk Assize no less than thirteen gipsies were executed. This barbarity was finally swept away by a repeal of the statutes by King George III.

Before James V fell foul of the gipsies, some of them had reached positions of some power. The most remarkable of these was the celebrated Johnie Faw (or Faa), Earl of Little Egypt, who (as the old ballad tells us) carried off the Countess of Cassillis, helped by his fifteen trusty followers.

' " Oh come with me," says Johnie Faa ;
" Oh come with me, my dearie ;
For I vow and swear by the hilt of my sword,
That your lord shall nae mair come near ye." '

The Faas are the royal line of gipsies in Scotland, and Queen Esther – whose coronation I have described earlier – was of their blood.

In spite of prison and death the Romany Folk have kept their freedom, retained their ancient language, and remained faithful to their ancient chiefs or kings. But to-day they are gradually dying out. No longer are they outlawed or banished, and, though a race apart, they enjoy all the rights and privileges of any British citizen. Yet the laws of economics are harder to evade than the edicts of mediæval tyrants, and the times are against the Romany Folk. Commons and open spaces are enclosed as never before, the old country fairs which brought so much of their revenue are passing away, and even the skilled trades at which they once excelled are being snuffed out by modern progress. Cheap German hardware has killed the craft of the tinker, and American plywood chair seats have cut out the older rush seats which the gipsy chair-mender could so skilfully repair. Foreign crockery is now so cheap that there is no longer a living to be made in riveting plates and dishes, while the ever-increasing number of 'stores' has ruined the gipsy trade in basket-work chairs, etc. The motor-car has disposed of the 'hay-motor,' so that the horse-dealing in which the Romany Folk once excelled is no longer profitable; and even the harmless fortune-telling of the gipsy-wife is illegal, and she liable to fine or imprisonment, which in my judgment is grossly unjust in view of the large amount of license permitted to smart society clairvoyantes and crystal-gazers and their kind, who rake in more pounds in a day than the humble gipsy pennies.

Even the roundabouts, once the profitable preserve of the richer Romanies, are ceasing to attract a generation bored to tears with too much amusement.

It is remarkable that the gipsies have kept their ancient language and their racial purity without even the bond of a common religion. That, and that alone, accounts for the survival of the Jews as a separate race, but the gipsies have no common religion and their sole bond is an overpowering love of freedom and the open air. They have other good points. They may beg, but they would rather sleep under a hedge on a wet night than accept the hospitality of the casual wards. They are very thrifty – in spite of their drinking habits – and when they get into trouble with the police, always seem to have the amount of their fine.

As a Magistrate I have noticed this, and that the most wretched, ragged, miserable-looking objects imaginable will produce several Treasury notes when necessary. They are willing to work hard too, provided it is not too long ; but they will not enter regular employment. Pea-picking, fruit-picking and hop-picking carry them through the summer months, with hawking in the winter. They visit our great fruit-growing districts regularly, and the farmers provide camping grounds to which the same families return year after year. I have met many of them in Kent, Hampshire and Cambridgeshire during the fruit season. They were picking strawberries for a farmer at Titchfield one summer ; the weather was dry and the crop was short, so they demanded an extra penny a basket. Pickers are usually paid so much a basket for all fruit picked. The farmer refused to pay any more and the Romany Folk left the job. Two days later the farmer endeavoured to get them back, but they refused unless the advance was paid. The farmer pleaded that the amount demanded would involve him in a loss on his crop. The gipsies said that the fruit still remaining in the field was worth at least £20. "Nonsense," replied the farmer, "it's not worth £10." "I'll give you £10," said the gipsy. "Done," said the farmer, never thinking they had the money. But they had, and paid cash down; and if my Romany friends told me the truth they made a profit on the deal. Certainly I saw them picking the fruit they had purchased, which they were astute enough to sell before picking to a jam manufacturer, whose motor-lorry was collecting fruit in the district.

It is interesting to see them trekking, as they move on from one district to another. They have a curious method of marking their route at road forks and crossroads, so that those who follow will know which way their leaders have gone. A simple cross is scratched on the road, and the long arm points the way. Sometimes a long and a short stick perform the same service. An old gipsy put it in their ancient language: "Duveleste avo. Mandy's kaired my patteran adjusta chairuses where a drum jals atut the waver." (Translation: "God bless you, yes. Many a time I have marked my sign where the roads cross.")

I have no space to go into the fascinating subject of the gipsy language. It is said that men of their race can converse with each other whether they are Spanish, French, German or British. My own knowledge of the language is too slight to verify this, though I have met and tried to converse with the Romany Folk in many lands, and particularly in the Balkan States, Spain,

France and England; but I have noticed certain words common to all. In considering their language, however, it is important to remember that it is purely a spoken dialect of a semi-illiterate people; there are no grammars and no books, and pronunciation varies even among gipsies in the same country. It is probable that the gipsy tongue, as it exists in England to-day, is merely a corrupt dialect, containing few inflections and mixed (as it certainly is) with a greater or less number of English words. It conforms to English usage in the arrangement of the sentences. But there is, or was till recently, an older, purer tongue. This is the Deep, or ' Old Dialect,' known to only a few old folk to-day, containing many more inflections and idioms and closely resembling the Continental gipsy dialects. It is a fascinating but perplexing task to ' try out ' the gipsy language (as printed in such books as Smart and Crofton's *English-Gipsy Vocabulary*, Borrow, or Morwood) on a live gipsy to-day. Only the old folk are any good, and they are often pleased to help, provided always you will cross their palm with silver in the traditional manner. It is useless, however, to expect more than words and their pronunciation. All my efforts to get at the grammar and construction have been useless, for the gipsies themselves do not know. But they do know the joy of the open air, the lilt of the bird in the tree, the song of the wind in the heath; and they will say with Eliza Cook :

'Our fire on the turf, and our tent 'neath a tree –
Carousing by moonlight, how merry are we!
Let the lord boast his castle, the baron his hall,
But the home of the gipsy is the widest of all.
We may shout o'er our cups, and laugh loud as we will
Till echo rings back from wood, welkin and hill;
No joy seems to us like the joys that are lent
To the wanderer's life, and the gipsy's tent.'

GIPSIES AT A FAIR.

ROMANY CHILDREN.

To face p. 194.

OCTOBER

*FOX-HUNTING, ETC.
HALLOWE'EN*

THOUGH scarcely to be classed as folklore, Fox-Hunting is such a picturesque survival that no work devoted to the 'Colour of the Countryside' can omit all reference to the Meet. Though hunting is a thing of the past for me, I have many thrilling memories of the chase – the thud of hoofs on close-cropped turf, the sting of keen air on face and nostrils, the note of the huntsman's horn, the music of the pack, the tightening of nerves and muscles when a stiff fence is tackled and the exultation when it is surmounted and we are still all in one piece.

And the Kill – barbaric no doubt, but unforgettable when one has risked life and limb to get there. A solid mass of yelping hounds hurling themselves on their weary quarry, the dash of the whipper-in to save the corpse as he beats back the frantic pack with butt end of whip. Then he raises the dead beast and quickly severs its brush, mask and pads to become valued trophies to some follower, and the mutilated carcase is flung back to the waiting pack. Another surge of eager dappled bodies and in a moment no vestige of the fox remains save a heavy odour of musk – once smelt never forgotten. But I think the custom of 'blooding' the youngest follower with the gory severed members of the dead beast might well be given up. Hunting in these democratic days is an obvious anachronism (like the swivels still worn in the belts of the

Yeomen of the Guard to carry muskets which have been out of service for three centuries) – a survival of the feudal period; and if we desire it to continue we would do well not to flout the opinions of the majority – a lesson the staghunters of Exmoor have yet to learn. In spite of its economic absurdity, Fox-Hunting is a real sport (which the modern massacre of hand-fed pheasants by paunchy profiteers certainly is not), and adds a delightful dash of colour to the drear November countryside. How charmingly the red coats, sleek horses and dappled pack blend with the browns and greens of our winter landscape, and what a feast for the eye is afforded by a Meet on the turf in front of some Tudor Manor-house, or in the market square of some country village. Even to-day, when Hunt Committees are anxious to conciliate farmers, we hear of occasional damage to crops by huntsmen, and to chickens by foxes; though probably the injury to agriculture by the hunting interest is not a twentieth of that caused by overstocking with game. If space permitted, it would be a fascinating task to sketch the whole story of the chase, from primitive days when it was the birthright of every inhabitant, on to the ages of tyranny when kings and nobles usurped to their own order what had once been the common heritage of mankind. The nadir was reached under Norman rule, when the killing of the king's deer was punishable by death, or what was worse – perpetual blindness; and for a man to drive the deer out of his own field was a crime. But even the Norman bandit-kings could not for ever maintain such harsh penalties, and after the death of William the Red had pointed an obvious moral, we find a slow progressive improvement. But it was slow. Mr. J. C. Cox, F.S.A., in his masterly book on the Royal Forests of England, quotes a case in 1336 when Thomas Bret, Vicar of Scalby in Pickering Forest, and four others were fined for making a fence of small thorns to guard their sheep from the fox.

A very interesting and highly picturesque survival in this connection are the Verderers' Courts, which formerly existed in many districts. I know of two to-day – one at Lyndhurst, and the Speech House in the Forest of Dean. Both are still working, and the Verderers who compose the court have magisterial powers in their own sphere. Both of these old courts are decorated with the antlers of deer, and that at Lyndhurst boasts a stirrup iron said to be that used by King William Rufus when he was killed, but probably of much later date. The lover of the past will hear many quaint phrases in the Verderers' Court – Vert and Venison, Turbery, the agistment of cattle and the pannage of swine. He will find too that in the New Forest many valuable rights are still possessed by the freeholders, such as the power to turn out cattle and the right to cut turves for fuel. The latter is a much prized privilege, known as Turbery, and all the old Forest cottages and houses have this right, the quantity being decided by the size of the fireplaces. That is why the large old-fashioned open fireplaces still remain in old houses which have otherwise been completely modernised, and when a cottage is burned down or otherwise destroyed the fireplace remains. We sometimes see an old brick fireplace standing gaunt and empty in a deserted garden, but it is not useless – it retains for its owner the right of Turbery.

These benefits are, of course, similar to the old common rights, which are somewhat outside the scope of this book, but one of them, which I discovered in Wiltshire, carries a most interesting bit of folklore which I must include. I refer to the lovely old town of Malmesbury, where the beautiful Common still known as the King's Heath and an estate near Norton were given to the freemen by King Athelstan. These rights are really valuable, since each freeman is entitled to an allotment and every capital burgess to a plot of eight to fifteen acres. Those believing themselves to be entitled to benefit must make their claim in person to the court, which is a thousand years old this year (1930). If a man is successful a most interesting little ceremony takes place. The President calls him up to the table, and solemnly hands him a piece of turf and a twig of a tree, using these words :

> " Turf and twig I give to thee,
> The same as King Athelstan
> Gave to me."

These old common rights are now restricted to

the son of a commoner or the husband of a commoner's daughter, and commoners to benefit must live within the walls of the town. Mr. Hutton tells of an old bedridden man who would not go into the 'House' for fear of losing his common rights. "King Athelstan hath kept I all my life, King Athelstan shall keep I till I die."

October 31st

HALLOWE'EN

THE correct name of this popular festival is All Hallow's Eve, being the vigil of All Hallow's Day, November 1st. The leading idea connected with this festival is that it is the time beyond all others when supernatural influences prevail. I can remember its celebration in Northern Ireland, where the custom had a strong hold.

The girls and boys, but especially the former, endeavour to discover the probable course of their love affairs by means of nuts. Two are placed side by side on the bars of the grate to represent a pair of lovers, or if a girl has two suitors she will name each nut after one of them, and place one in the middle to represent herself. If the nut cracks or jumps away the lover it represents will prove unfaithful, but if two blaze together they will be married. Mr. Charles Graydon has put the idea into verse charmingly thus :

> 'These glowing nuts are emblems true
> Of what in human life we view ;
> The ill-matched couple fret and fume,
> And thus in strife themselves consume.
> Or from each other wildly start
> And with a noise for ever part.
> But see the happy happy pair,
> Of genuine love and truth sincere,
> With mutual fondness while they burn,
> Still to each other kindly turn ;

> And as the vital sparks decay
> Together gently sink away;
> Till life's fierce ordeal being past,
> Their mingled ashes rest at last.'

These customs have given the date the name of 'Nutcrack Night' and are widely spread over the British Isles, as Burns wrote in his 'Hallowe'en':

> 'The auld guidwife's well-hoordit nits
> Are round and round divided,
> And mony lads and lasses' fates
> Are there that night decided.
> Some kindle, couthie, side by side
> And burn the gither trimly;
> Some start awa wi' saucy pride
> And jump out owre the chimly.'

Another favourite amusement of Hallowe'en is bobbing for apples floating in a tub of water. No use of hands is allowed, and much sport is provided by the efforts of the soaked competitors to secure the elusive prize.

NOVEMBER

BONFIRE NIGHT
LONDON'S LORD MAYOR'S SHOW, AND MAYORAL CUSTOMS GENERALLY
WROTH SILVER: ST. MARTIN'S DAY

November 5th

BONFIRE NIGHT: WITH SPECIAL REFERENCE TO THE CARNIVAL AT LEWES, SUSSEX

MOST, if not all, civilised nations have their great days of commemoration and rejoicing. The French National Fête on July 14th commemorates the Storming of the Bastille, which marked the fall of absolutism, while the 4th of the same month is celebrated with much vigour in the United States as Independence Day. Until the recent institution of Armistice Day on November 11th, no day had taken such hold on the imagination of the English People as the Fifth of November – Guy Fawkes' Day. The dramatic circumstances in which this plot was frustrated, as well as the Englishman's traditional hatred of foreign despotism, whether civil or religious, have served to ensure the observance of the day down to our own time, and bonfires are kindled in every village in England save one. That is Scotton, near Knaresborough, Yorks, close to the Hall where the arch-scoundrel lived as a boy, and you must not guy Guy there, for his ghost is still active. To-day the celebrations have lost their political or religious character and are merely a jolly carnival of youth, with two exceptions – and they are significant.

The greatest of all bonfire celebrations is held

at Lewes, Sussex, and in 1929 no less than one hundred thousand torches were burned during the evening. Here the observance partakes of the character of a religious ceremony, and there is much of the same spirit in another old-world town – Bridgwater, Somerset. It is very interesting to notice that these two ancient market towns, fully two hundred miles apart, whose history, outlook and character are widely different, are yet unique in all England in maintaining the political and religious character of this celebration and regard it as a great day of rejoicing and thanks to Almighty God for deliverance from peril. In the awful days of the Marian persecution seventeen poor creatures were burned alive in Lewes Market Place. Two were men of substance, a Brighton brewer and a Warbleton ironmaster; and the rest were humble village folk, whose sole crime had been to read the Bible for themselves and live simple godly lives. On June 22nd, 1557, ten poor villagers were burned alive in one gigantic blaze, six men and four women, and the memory of this dreadful deed will never be forgotten in Sussex. In Bridgwater it is the recollection of the Bloody Assize that gives zest to the 'Festival of Deliverance.' I have no space to tell the tragic story of that assize, by which Judge Jeffreys won the Great Seal from his master, but I well know the shabby little public house, down Wapping way, where the Judge was found by an angry London crowd and hurried to the Tower; and when in Barbadoes I found another interesting memory of his evil days. Noticing some white men diving with niggers for small coins in Bridgetown Harbour, I asked my friend – a local planter – how it was that white men so forgot the rigid colour line. " Oh," he replied, " they are ' red-legs,' or ' mean whites.' Their forefathers were sent here as slaves after Sedgemoor, and their hardships in the cane plantations so broke their spirits that when they regained their freedom, they could never win back full status as white men."

But I must return to the Fifth at Lewes. There are no less than six Bonfire Societies – the Commercial Square, St. Anne's, Southover, Borough and Cliffe being the most important. The most historic is the Cliffe Society, which still uses a real eighteenth-century ' No Popery ' Banner, and figures of Pope Paul IV (who is regarded by the Bonfire Boys as the real author of the Gunpowder Plot) and Guy Fawkes are carried in procession and burned. The effigies are filled with fireworks which explode when the fire reaches them. This is a decided improvement on the methods of a century ago, when the papal belly was filled with live tom-cats, and the noise which the unfortunate creatures made when the flames reached them caused the utmost merriment among the crowd.

In spite of the bonfires and fireworks, the Lewes celebrations have a definitely religious character. The processions start at about 5.30 p.m., and each society in turn marches to the War Memorial

BONFIRE DAY AT LEWES.
A CURIOUS SURVIVAL.

To face p. 206.

and lays a wreath in memory of the Bonfire Boys who gave their lives for their country in the war; afterwards there is a short address by a clergyman and a hymn.

The Cliffe Society, which alone retains the historic character of the proceedings, still uses the old eighteenth-century Bonfire Prayers.

Torchlight processions continue all the evening, and the effect is most impressive as the long lines of flaming torches wind through the narrow, steep streets of the ancient town. The final scene takes place about 11.15 p.m., when each Society marches to an open place outside the town and burns its effigies. Formerly the bonfires took place in the streets of the town, and lighted tar-barrels were rolled through the main thoroughfares. Owing to the obvious danger of serious fires, the police attempted on a number of occasions to stop the celebrations altogether. This led to serious rioting, especially in 1847 when the stupid conduct of Lord Chichester caused very serious casualties; and finally the authorities gave way on condition that the Bonfire Boys themselves were responsible for the good behaviour of the crowds. This has worked well, and though there is still plenty of fun, there is no horseplay or damage, and the great festival is well worth seeing. The Cliffe Bonfire Society, by the way, is the oldest institution of its kind in the world, and is still going strong.

November 9th

LONDON'S LORD MAYOR'S SHOW, AND MAYORAL CUSTOMS GENERALLY

THOUGH shorn of its antique pageantry, and bereft of much of its former significance, the procession which passes through London to Westminster every 9th of November is now simply styled 'The Lord Mayor's Show.'

In these utilitarian days we hear many complaints of the hindrance to traffic caused by this function, and suggestions for its abolition are not infrequent. From the point of view of all who love the lore of the past it is to be hoped that it may long continue, and indeed so long as similar processions are permitted (e.g. Socialist May Day, Royal and Ecclesiastical pageants) it would be intolerable if London's greatest citizen should not be permitted to receive the homage of the people.

There is a tendency nowadays in some quarters to sneer at civic functionaries, particularly at those who possess exalted rank without owning the qualities which ought to go with it. The fat little grocer mayor of some small town has long been a popular butt for the comic artist and the humorous papers (and it is true that occasionally the small shopkeeper becomes a most divertingly pompous mayor), but it should never be forgotten that even in the smallest town it is impossible to reach the position of Chief Magistrate without years of honourable, unpaid public service, and I believe there is no country in the world which can boast such a fine body of honest, patriotic, intelligent and progressive civic fathers as England. In view of this I maintain that the Lord Mayor of London, as Chief Magistrate of the greatest city in the world, holds an office of more real dignity and importance than many kings, and, moreover, has attained to it by character and worth instead of the mere accident of birth. Nor should it be forgotten that in the ages of tyranny the Lord Mayor was a most powerful defender of the liberties of the citizens, and established the mighty principle that kings had no jurisdiction within the City limits. That is why a Royal procession has to halt at Temple Bar and ask permission to enter the City; and no troops can pass through without the Lord Mayor's sanction. Further, the special password, which is changed nightly, and without which nobody may enter the Tower after dark, is forwarded to both the King and the Lord Mayor. When we recall that the Tower was the fortress of mediæval absolutism, just as was the Bastille under the *ancien régime* in France, we shall appreciate how far the Lord Mayor had established his position even then. This is further illustrated by the right of the City Corporation to carry its maces upright (instead of sloped) a sign that they are ready for action and a symbol of sovereignty. The Lord Mayor also occupies an important position near the King at the Coronation, and wears the coronation robes of a Peer.

But I must return to the Lord Mayor's procession. There are two remarkable wooden figures, known as Gog and Magog, which ordinarily inhabit the Guildhall, but are occasionally carried in the procession. They have an interesting history. The old legend relates that Gogmagog was a giant who was slain by Corineus, a gigantic follower of Brutus of Troy, who invaded England. Of the two figures in the Guildhall the elder is Brutus and the younger is Corineus. The figures were carved in 1707, but earlier specimens are recorded. Two of them were placed on London Bridge when Philip and Mary made their entry into the City.

November 11th

WROTH SILVER: ST. MARTIN'S DAY

RECENT legislation in connection with real estate will ultimately sweep away a large number of semi-obsolete feudal and manorial rights. Doubtless most of these are now out of date and vexatious, and their disappearance will cause no regret. There is, however, one highly picturesque old custom which hitherto has been regularly observed half an hour before sunrise on the morning of November 11th.

This is the Payment of Wroth Silver at Knightlow Cross, Stretton on Dunsmore, near Rugby, Warwickshire. At the northern extremity of the village, in a field near the Holyhead Road, stands a stone, the only part remaining of the ancient Knightlow Cross. The Lord of the Manor is the Duke of Buccleuch, and his steward attends here about 7 a.m. He reads a notice requiring the payment and giving the names of those responsible, and proclaims that ' in default of payment the forfeit would be 20/– for every penny, and a white bull with red ears and a red nose.' Thus, though the payments required from the persons and parishes are trifling, ranging from 1*d.* to 2/3½, the penalty for non-payment is substantial, particularly as it would prove exceedingly difficult to obtain a bull marked as prescribed. The money is thrown into a cavity of the stone and taken out by the steward.

After the ceremony the actors in the scene – usually about forty – are entertained to breakfast at the expense of the Duke, who certainly does not make a profit on the transaction. The origin of the custom is somewhat of a mystery, but it is probable that these payments were wayleaves for the use of roads in the manor by the tenants' cattle.

If this picturesque old custom should be discontinued as a result of the legislation mentioned, English folklore will certainly be the poorer.

DECEMBER

*THE CHRISTMAS CEREMONY AT KILLIN,
  PERTHSHIRE*
THE BOY BISHOP OF BERDEN, ESSEX
THE CHRISTMAS MUMMERS

THE CHRISTMAS CEREMONY AT KILLIN, PERTHSHIRE

ON pages 164–6 I have fully described the quaint and interesting custom of Rush-bearing at Grasmere, but an even more curious observance is still maintained at Killin, Perthshire, at Christmas. I know this charming village well, but as my visits thither have been during summer – I hate the Highlands in mid-winter – I have not seen its Christmas custom. I therefore refer to Dr. George C. Williamson's book, *Curious Survivals* (Herbert Jenkins, Ltd.).

The church, as the village name implies, is dedicated to St. Fillan, and preserved near it are a number of stones, large and small, which are said to have been used by the saint for curing sprains and other ills in both men and cattle. The stones are water-worn, and were obviously derived from the rushing torrent which still roars through the village into Loch Tay. It appears that the saint employed the stones to stroke the injured parts, thus anticipating modern massage. They are now preserved within a grated niche, on the inner side of the eastern gable of an old meal mill belonging to the Marquess of Breadalbane. Every Christmas these stones are taken out, and placed on a bed of ' Water-carried straws, and rushes uncut by the hand of man.' The rushes are pulled, not cut, and the stones are placed on them. The ceremony is said to date from mediæval times.

December 7th

THE BOY BISHOP OF BERDEN, ESSEX

ALTHOUGH the county of Essex actually runs into London, and dingy suburbs of the great metropolis straggle far into its broad acres, it still contains many old villages strangely remote from main roads and railways. One of these is Berden, about six miles north of Bishop's Stortford, a peaceful spot, remote from change, where old inhabitants can trace their descent back through many generations of ancestors who lived in the same place. There is a family living at King Harold's farm who are descended from that monarch, and another who are descended from the famous Turpin line.

Here on December 7th the old-world ceremony of ' Enthroning the Boy Bishop ' is still kept up, as it has been – with some gaps – from mediæval times. The Boy Bishop is a choir-boy about eight years old. He wears full vestments and carries a crozier, and is elected by his fellow choristers. It is interesting to mention that the Boy Bishop in 1926 was Harold Goodman, a descendant of King Harold ; and in 1929 it was Frankie Turpin, a descendant from the celebrated Turpin family. Dick Turpin was born at the Crown Inn, Hempstead, about twelve miles north-east of Berden, and various places associated with his story are still pointed out by the villagers. It is a remarkable instance of the uncertainty of fame that though William

Harvey, the famous surgeon who discovered the circulation of the blood, was also born at Hempstead, nobody whom I met there had ever heard of him, but everybody knew about the sordid scoundrel who became a 'hero' through Ainsworth's masterly novel.

THE CHRISTMAS MUMMERS

> 'Tu shorten winter's zadness
> Zee where the folk with gladness
> Disguised all be comen
> Right wantonly a mummen.'
>
> (*Old Song.*)

EIGHT hundred years old and still running! This surely is a record for the English stage, and beats *Charley's Aunt*, *Chu Chin Chow* and *Journey's End*. But the very interesting old folk drama still performed in various villages in Hampshire has been continually carried on every Christmas since the time of the Crusades, and I am glad to say seems likely to last for many years yet.

I have kept in close touch during the last few years with two troupes, the Overton and the Longparish Mummers, both of which perform regularly. I have also seen during the same period similar exhibitions at Burghclere and North Waltham in north Hampshire, and Southwick and Bursledon in the south of the same county.

Though now a rare and remarkable relic of the past, the Mummers were once to be found in nearly every village all over England. During the last fifty or sixty years they could have been seen in most of the country places from Cornwall to Sussex, also in Gloucestershire, Wiltshire and Oxfordshire. It may be of interest to name a few of the villages, thus:

WARWICKSHIRE : Pillerton, Ilmington and Great Wolford.

GLOUCESTERSHIRE : Weston-sub-Edge, Sapperton, Icomb and Longborough.

OXFORDSHIRE : Waterstock, near Thame, Shipton under Wychwood and Leafield – villages near Burford ; Cuddesdon and Lower Heyford.

Though the play in its present form probably dates from about the time of the Crusades (say the twelfth century or thereabouts), there can be little doubt that it contains traces of pre-historic paganism. The central 'Act' of the drama is the fight between King (originally Saint) George and various opponents, all of whom are slain by him and subsequently restored to life by the Quack Doctor – the primæval medicine-man or witch-doctor. This may represent the death and resurrection of the old year, or relate to the slaying of a victim, at or before sowing time, to propitiate the Corn-god and ensure a good crop.

It is well known that wherever sowing occurs among primitive peoples in any part of the world, it is accompanied either by human sacrifice or by some ceremony which may be interpreted as the mitigation and vestige of some ancient sacrificial custom. In ancient Egypt it was the custom to sacrifice a fair-haired person to Osiris, the God of Harvest. The unfortunate was actually seized and slain in the harvest-field, while the reapers

MUMMERS AT LAVERSTOKE HOUSE.

THE FIGHT SCENE IN THE SNOW.
THE OVERTON MUMMERS.

knelt and prayed that the Corn-god might return the following year with his powers renewed and strengthened by the blood of the victim. A similar practice obtained in ancient Mexico, but there the victims were usually women, whose faces were painted yellow, while on their heads were placed tasselled caps to represent the ears of maize.

The Mummers' Play is of the utmost interest to all students of literature, because it is the *only* survival of the pre-Reformation folk drama. These plays were of two classes, afterwards confused, but at first kept quite distinct : *Mysteries*, or Scripture plays, and *Miracles*, plays dealing with the legends of the saints.

It is possible that the Miracle Play would never have come into being if the services of the Church had been carried on in the vernacular, instead of in Latin – a language 'not understood of the people.' As the words of the service could only be incomprehensible, it was necessary to enlighten the people by other methods. Hence the carvings in wood or stone, and marvellous west fronts like those of the Cathedrals of Wells, Lichfield and Salisbury, which were almost ' Bibles in stone.' Few, alas, of these splendours remain to us, but at Fairford in Gloucestershire I have seen a magnificent selection of genuine mediæval windows, which presented in pictures everything that the faithful had to believe. Presently there began at Christmas and Easter little tableaux in the churches, the Babe in His cradle at the former, the Body in the grave at the latter.

Then came the Mysteries – the first of the ethical dramas – followed by the Miracles and Moralities. *Everyman* is the best known of the latter. But the Reformation, by giving the Bible to the people, lessened the need for dramatic explanation; and the Renaissance, by opening up to the world the treasures of the classics, caused a mighty development of the secular stage. Thus the Tudor period saw a reconstruction of a literary stage founded on Greek and Roman models, which the genius of Shakespeare enormously developed and extended. Finally the Puritan period, with its hatred of pleasure in any form, and loathing of Popery in any attire, gave the *coup de grâce* to whatever remained of the ethical folk drama of mediævalism. The Mummers' Play alone remains, having survived the competition of both Seneca and Shakespeare, and the veto of Praise-God Barebones.

The Mummers' Play, then, is a genuine survival of mediævalism, and all the troupes with which I am acquainted are composed of ordinary unlettered working men. The words have been handed down by oral tradition from generation to generation, and in many cases the players follow a father and grandfather who also played his part. I was fortunate some years ago in obtaining from the leader of the Overton Troupe the full text of the play, which, he assured me, had never been committed to print before. I have also, from time to time, taken down verbatim snatches of the patter with which the various characters interlard their parts; and I have been careful to retain grammatical errors and obsolete phrases as far as possible. As I have already stated, I have seen at various times not less than half-a-dozen troupes, but the two with which I have kept the closest touch are the Overton and Longparish players; and the latter are by far the best-dressed troupe I have discovered.

Each of them still performs every Christmas, and for many years has regularly visited me and given a show on my lawn. Their usual procedure is to visit the houses of the principal residents in the daytime, and to entertain the customers at the various village inns after dark. Readers who wish to see this unique performance had better enquire at the inns, in whatever district they happen to be, and it is usually possible to ascertain their movements in this way.

The caste consists of at least six men, and sometimes eight. The leader, who introduces the party, is styled Father Christmas; King George (originally, of course, Saint George), the hero of the play, Bold Slasher (perhaps Bow Slasher), Quack Doctor, Lawyer, Valiant Soldier, Twing Twang (who also answers to the name of Little Johnnie Jack, an obvious reference to the mighty henchman of bold Robin Hood) and Rumour. When only six players are present, Lawyer and Valiant Soldier are omitted. Father Christmas sometimes duplicates the part of Rumour, and

Twing Twang of Little Johnnie Jack. This character is interesting, as he introduces himself thus :

" In comes I, little Johnnie Jack,
With my wife and family up my back.
One Doll is large, and the other small."

This is a possible reference to the Virgin and Child, and no doubt comes from some mediæval Mystery Play. The Rector of Brightwalton (Berks.) informs me that years ago the children of the village used to carry round decorated baskets at Christmas, containing a large and a small doll. This obviously has a similar origin.

The performance opens with a carol, all singing :

" God bless the master of this house,
With a gold chain round his neck ;
Oh, where his body sleeps or wakes
Lord send his soul to rest.
God bless the mistress of this house,
With a gold chain round her neck ;
Oh, where her body sleeps or wakes
Lord Jesus be her guide."

The first occasion on which I saw the Mummers' Play (some years ago) was at the house of Sir William W. Portal, Bt., F.S.A., and I am indebted to him for pointing out the significance of the words ' with a gold chain round his neck,' which date the verse before the middle of the

DETAIL OF 'LITTLE JOHNNIE JACK.'   CLOSE UP OF KING GEORGE.
Note dolls carried.

LONGPARISH MUMMERS.

To face p. 222.

fourteenth century. Up to the early part of that century the Lord of Castle or Manor-house was accustomed to take his meals in the Great Hall with his wife, his guests, retainers and servants, and wore a gold chain round his neck to distinguish him. In *The Vision of Piers Plowman*, published in 1362, William Langland made an attack on the wealth and luxury of the period, and one of the new-fangled customs of which he complains is that the Lord and Lady now take their meals in their private apartments instead of in the common hall ! The second reference to the ' Mistress of the House, with a gold chain round her neck ' was doubtless a later addition, the fourteenth century being one of male domination.

The various characters then introduce themselves in crude verse.

FATHER CHRISTMAS (who usually acts as stage-manager) says:

> " Here comes I, Father Christmas am I,
> Welcome — or welcome not ;
> I hope old Father Christmas
> Will never be forgot."

TWING-TWANG : " I hope he won't here."

FATHER CHRISTMAS *continues* :

> " Christmas comes but once a year,
> When it comes it brings good cheer ;
> With a pocket full of money
> And a cellar full of beer,

Roast beef, plum pudding and mince pie,
Who likes them any better than I?"

TWING-TWANG: "I do."

FATHER CHRISTMAS:

" I don't know that you do, my little feller;
I want room and room and acres of room;
After me comes King George with all his noble train."

CHORUS:

" He comes, he comes, our hero, hero comes.
Sound, oh, sound the trumpet, and beat,
Oh, beat the drum.
Loud along the shore the cannons roar,
Walk in King George, along the pretty shore."

Enter KING GEORGE:

" In comes I, King George, so bold, so grand
I do appear, with my old tribes and Britons
By my side; I am come to close this year.
Here is England's rights, here England's wrong,
Here is England's admirations.
When I pull out my old rusty rapier,
Is there a man before me can stand
That I can't knock him down
With my created [perhaps courageous] hand?"

The challenge is accepted by RUMOUR.

" I'll fight thee, King George,
Like a man of courage bold;
Let thy blood be ere so hot
I will quickly fetch it cold."

Father Christmas *replies*:

> "Ha, ha, my little feller, thee talks very bold,
> Like a good many more as I've been told."

King George:

> "Pull out thy rusty rapier, pull out thy sword and fight,
> Pull out thy purse and pay.
> My satisfaction is to have thee this night
> Before I go away."

Note the way in which the various protagonists brag like the heroes of Homer, or the chieftains of some Norse Saga ; this clearly points to this act having originated in the childhood of mankind, when the great ones of the tribe were swaggering, bragging, fighting men.

King George and Rumour fight with wooden swords (or sticks). The former, of course, kills his opponent, and repeats his challenge. It is taken up by Turkish Knight, this character dating from the time of the Crusades. Some of the Mummers call him 'Turkey Snipe,' an amusing instance of how words become corrupted through the centuries. Like all the others, he introduces himself in crude verse, thus :

> "In comes I, the Turkish Knight,
> Just come from the Turkish land to fight.
> Only me and seven more
> Fought and killed eleven score,
> Eleven score of gallant men,
> For the sake of George our King.

I will fight thee, King George,
Like a man of courage bold.
Let thy blood be ere so hot
I will quickly fetch it cold."[1]

KING GEORGE *replies* :

" In comes I, King George, that man of courage bold,
With my sword so valiant by my side.
I won ten thousand pounds [? crowns] in gold.
'Twas me that fought that fiery dragon
And brought him to his slaughter,
By those means, and many things,
I won the Queen of Egypt's daughter.
If any man dare enter this kitchen or hall,
I will cut off his head and kick it about like a football."

They fight, and King George, of course, slays his antagonist.

FATHER CHRISTMAS *then bursts out in lamentation for the dead men.*

" King George, King George, what hast thou done?
Thou hast ruined me by killing of my son.
Oh, is there a Doctor to be found
To heal or cure these two men
Which lie bleeding on the ground? "

[1] " In comes I, old foreign King,
With my broadsword I do swing.
Likewise I am the Turkey Snipe,
Just come over to old England to fight.
I'll fight thee, King George,
King George, thou man of courage bold,
And if thy blood runs hot
I'll quickly fetch it cold."

This alternative wording to the Turkish Knight's piece is used by the Longparish Troupe.

THE DOCTOR ADMINISTERS THE 'GOLDEN FROSTY DROP.'

THE DEAD MAN RESTORED.

THE LONGPARISH MUMMERS.

To face p. 226.

QUACK DOCTOR:

" Oh yes, oh yes, there is a Doctor to be found
    To heal or cure these dead or wounded young men
    Which lie bleeding on the ground."

FATHER CHRISTMAS:

" What canst thou cure, Doctor ? "

QUACK DOCTOR:

" I can cure the itch, the stitch, the palsy and the gout,
    The region[1] pain both in and out [etc.].

    I have a small bottle in the waist-band of my belt
    Called the ' Golden Frosty Drop.'
    Two drops of that will fetch thy two sons
    To life again."

FATHER CHRISTMAS : " Try thy skill, Doctor."

QUACK DOCTOR:

    " A little to the eye, a little to the thigh,
    A little to the string bone of the heart.
    Rise up, you men, and try to stand
    And see the time of day.
    After you have done, put out your tongue
    And let's hear what you can say."

The Doctor then administers a dose from the bottle to the men lying on the ground. Some troupes make this an opportunity for fooling with a gigantic pill, or horrible black draught.

The ' dead men' rise and address the Doctor : " Home we go."

[1] Probably raging.

FATHER CHRISTMAS :

> " Well done, my little man,
> Thee bisn't like those half quack doctors ;
> Thee does the work. Will thee have the money now
> Or stop till thee gets it ? "

This sometimes concludes the performance, but when there is good financial support from the audience, an extra ' turn ' is introduced.

This consists of a scene between Valiant Soldier and Lawyer. It amusingly illustrates the popular idea of the legal profession in the fourteenth century and after. It will be remembered that in the Peasants' Revolt of 1381 the insurgents murdered every lawyer they could get hold of, as they were regarded as the willing instruments of tyranny and were particularly hated for the part they bore in the oppression of the labourers through the Statute of Labourers, etc.

In this scene, Valiant Soldier, having challenged King George, is wounded but not killed. While lying on the ground he says :

> " Oh dear, is there a lawyer to be had,
> As I sets here so very bad ? "

THE LAWYER *proffers his services, but first asks how much money he has.*

VALIANT SOLDIER :

> " Two shillings in money, one wife, one son ;
> When that's set down the Will is done. "

The Lawyer *demands a fee of ten guineas and a glass of grog.*

Valiant Soldier : " Thee be well aware I can't pay that."

Lawyer : " I am supposed to take your cap, boots and clothes, and if I had a lawyer's rights, I should take your life as well ! "

This goes on to greater or less length, till the last character introduces the collection-box with these words :

" Here comes I, little Devil Doubt !
If you don't give me money I'll sweep you all out.
Money I want, and money I crave,
If you don't give money, I'll sweep 'ee all to the grave.
Gentlemen and ladies, since our sport is ended
Our box must now be recommended.
Our box would speak if it had a tongue,
Nine or ten shillings will do it no harm,
All silver and no brass."

The performance then closes with the well-known carol (all singing) :

" God rest 'ee merry gentlemen,
Let nothing 'ee dismay" [etc.].

The wording varies somewhat between different troupes, and some are in the habit of introducing humorous patter between the lines. Some of it is worth reproducing here. Thus, after the Opening Chorus, Father Christmas addresses the company thus :

"In comes I, all hind before, I come first to open the door. Make room for me and all my jolly family. I have a large family at the door, I hatched them in a magpie's nest, bred them up in a saw-pit, and they haven't had but one crust of bread this last fortnight, and that I ate myself! Don't you think me a good ol' dummin (old woman)?

"I went down a long, broad, short, narrow lane, and there I met a pig-stye tied to an elder bush, built with apple-dumplings and thatched with pancakes. I knocked at the maid, and the door came out, and she asked me if I could eat a glass of beer, and drink a crust of bread and cheese. I said, 'No, thank you, but yes, if you please'!"

Some of the other characters introduce themselves in words which are worth recalling. I have already mentioned Little Johnnie Jack; his twin, Twing-Twang, uses these words :

"In comes I, but I'm not come yet,
With my great head but little wits ;
My head is big, my wits are small ;
I'm just come in to please you all."

The name has an obvious reference to the sound of a bow-string, and reminds us of Bold Robin Hood, as well as his lusty henchman.

Rumour *says* :

> " In comes I, Rumour, Rumour is my name.
> I am come to show you merry sports
> To help pass away the winter time.
> Here is old activity, new activity,
> Such activity you never saw before."

In some troupes we find a semi-comic character called Belzybob, which is doubtless our old friend Beelzebub from the Mystery and Miracle plays. In these, as in nearly all folklore, the Prince of Darkness is more fool than knave. If he throws a rock or mountain at anybody he usually misses, and he is such an idiot that any old woman coming out of her cottage with a candle at midnight will frighten him away, as he will mistake the most wretched of farthing dips for the sunrise. He introduces himself with these words :

> " In comes I, Beelzebub,
> Over my shoulder I carries my club,
> In my hand my dripping-pan ;
> Don't you think I'm a jolly old man ? "

It is difficult to discover the origin of the club, but the dripping-pan is possibly a memory of some mediæval miracle-play in which the Enemy of Souls is equipped with the dripping-pan in order to baste them.

In the wording of this play I have introduced elements from a number of different versions, but mainly from those of Hampshire.

I have obtained the words from the lips of the players themselves, and here it may perhaps not be out of place to mention that the rural working man (and especially the agricultural labourer) is to-day our only living link with the mind and thought of yesterday. I commenced my study of the subject a quarter of a century ago, and in the interval many grand old veterans have gone over to the great majority. In a few years the Mummers' Play will have ceased to be. It cannot compete with the glamour of the towns, the cinema, the wireless, and the popular Press. Again and again I have seen a village troupe die out ; the ' old 'uns ' are gone, and the young fellows are ashamed to take it up. One thing, however, may change their view, and that is publicity. Some years ago I gave a Mummers' Troupe a big ' write-up ' in the Press, and as a result a couple of important London dailies sent their photographers down next year, interviewed them, photographed them, and made them the heroes of the village.

As a direct result (I believe) of that, an almost moribund troupe in another village woke to new enthusiasm and is still going strong.

If this book, by increasing interest in the subject, encourages more young fellows to give up their scanty leisure to learn their parts, and make up their complicated and interesting dresses, I shall be glad. The dresses, by the way, are worth mentioning. A feature common to all is that the faces are hidden by strips of paper ; at

THE OVERTON MUMMERS.

To face p. 232.

one time they were blackened. This habit of going about in bands in disguise led to some regrettable incidents, as in Tudor times bands of wandering rogues used the Mummer's attire to assist criminal ends, and drastic orders against this can be discovered by the historically minded. Most of the Mummers to-day make up their costumes from wallpapers, but several troupes, including Longparish, use some kind of thin, coloured material, which – not having been apprenticed to the drapery trade – I am unable to identify. Mr. Edmund Barber has pointed out that the Overton Mummers' headdresses resemble a Norman helmet, and the clothing a coat of mail. I gave him a set of photos of this troupe, and he was able considerably to increase my store of knowledge on this little known subject. Another writer – whom I fancy I once met when a-mumming – was the late Mr. R. J. E. Tiddy, whose book *The Mummers' Play* was published after his death. It is tragic that such a brilliant man should have been killed in the war. He was an enthusiast on the subject: and if any of my readers require more information, especially as regards the different versions of the Play, I can recommend it to them.

I fancy I have succeeded in increasing interest by numerous photographs which I have sent to the Press, and only last Christmas (1929) the leader of one troupe told me that a party had come by motor a hundred miles to see the play, as the result of some photographs they had seen in a paper the previous Christmas. This book, by giving fuller information on the subject than has yet appeared, will, I hope, considerably extend the circle to whom the Mummers appeal.

234

INDEX

INDEX

A

ABBOT'S BROMLEY, 170–2
Alfred, King, 50–1
All Fools' Day, 61
Aram, Eugene, 160–1
Ascension Day, 81
Ashbourne football, 21
Athelstan, King, 199
Atherstone, football, 21

B

BAAL WORSHIP, 69, 99–100
Bampton (Oxon), 105, 110
Barwick festival, 71
Bath, 89
Bath curse, 91
Bean Dance, 106
Beefeaters, 55, 121
Beltane fires, 69
Belzybob, 231
Benediction of the Throat, 18
Berden, Boy Bishop of, 215
Biddenden Dole and Twins, 56–8
Binding Monday, 42
Bladud, 89
Bonfire Night, 204–10
Bonfires, 69, 204–10
Bread and cheese lands, 57
Bridgwater Bonfires, 205
Brightwalton, 222
Buxton, 91
Bromley (Kent) festival, 70

C

CERNE GIANT, 134
Chaulkhurst sisters, 57
Chester-le-Street, 21
Christmas, Father, 221 *et seq.*
Christmas Mummers. *See* MUMMERS
City Companies, 151–2
Connaught bonfires, 69
Cornet, 127–8
Cowfold Dole, 40
Crawls, 31

D

DEERMEN, DANCE OF THE, 170–2
Derg, Lough, 97–9
Devil Doubt, 229
Doles, 28–40
Doon, Well of, 97
Druids, 130, 137–45, 160
Drumming Well, 96
Dunmow Flitch festival, 115–17

E

EASTER CUSTOMS, 52–60
Eggs, Easter, 58
Eggs, rolling, 58
Ellington Dole, 37

F

FAA OR FAW, 183, 189
Fairs, 173–80

Feet-washing, 54–5
Feoffement (Court), 45
First Foot, 2
Fish trade, 12
Flodden Field, 124
Flora, 77
Football, street, 21
Fox-Hunting, 196–8
Furmity, 179
Furry Dance, Helston, 76–9, 178

G

GAUNT, JOHN OF, 43
Gipsies, 181–94
Glastonbury, 86
Gowk, 61
Grasmere, 164–6
Guizer, Jarl, 10–14
Gunpowder Plot, 206
Guy Fawkes' Day, 204–7

H

HALLOWE'EN, 201–2
Harting, South, 118
Harvest lore, 160–3, 218
Haunted Pools, 88
Hawick ceremony, 124–8
Healing Wells, 89 *et seq.*
Helestone, 79, 133
Hellstone (legends), 79
Herrick, Robt., 24
Highland Games, 167
Hocktide at Hungerford, 42–51
Holly Boy, 24
Hood. *See* ROBIN HOOD
Horn-blowing, 44
Hunting, 105, 196–8

I

ISLAM (self-torture in), 100–1
Ivy Girl, 24

J

JACK IN THE GREEN, 74, 121
James IV, 188
James V, 189
Jesters, 121

K

KERN BABY, 161
Keyne St., 92
Killin, 165, 214
Klyack, 162
Knutsford Festival, 71, 119–21

L

LERWICK, 10
Lewes (Bonfire festival), 204–7
Lingam. *See* PHALLUS
Longparish Mummers, 217 *et seq.*
Lupercalia, 23
Lyndhurst, 198

M

MABELLA, LADY, 30 *et seq.*
Magdalen College, 68
Malmesbury (Common rights), 199
Manorial customs, 45
Mark's Day, St., 63
Martin's Day, St., 211
Marvin, Lady, 37
Maundy ceremony, 54–6
May Day, 66–75, 111, 120
Mayor, Lord, 208–10
May-poles, 71–2
May Queens, 70–74, 121
Meriden, 159
Midsummer Morning, 74
Minehead festival, 71
Morris Dancers, 102–14

Mummers, 102, 158, 217–33
Munster bonfires, 69
Mystery plays, 155, 219, 220, 231

N

November, Fifth of, 204–7
Nutcrack Night, 202

O

Odin, 126–7
Olney festival, 19
Overton Mummers, 217 *et seq.*
Oxford, 68, 107

P

Pancake Customs, 19–20
Passover, 53
Patrick, St., 87, 97–9
Pepys, 23
Pewsey, 129
Phallic symbolism, 46, 73, 75, 78, 85, 95, 99, 133–4
Poisson d'Avril, 62
Preston egg-rolling, 58
Punch and Judy, 154–8
Purse-strings, 55

Q

Quack Doctor, 221 *et seq.*

R

Riding the Marches, 124–8
Robin Hood, 76, 109, 171, 221, 230
Romany people, 181–94
Rush-bearing, 164–6

S

Samer, 186
Sanding, 119–20

Scotton, 204
Seville (festival), 52
Seville (Santa Semana), 52
Shoeing the Colt, 48
Shrove Tuesday, 19, 21
Southampton, 68
Speech House, 198
St. Blaise, 18
St. Cross, 38–9
Stocks, 178
Stonehenge, 79, 87, 126, 130–4, 137–45
Sun-worship, 74, 85, 100, 130–4, 137, 143
Swan Upping, 152

T

Teribus, 126
Thorn, holy, 86
Tichborne Dole, 28
Tissington, Well-Dressing at, 80–4
Town Criers, 44, 129
Tree-worship, 85–6, 104
Turbery, 199
Turkey Snipe, 225–6
Turkish Knight, 225
Turpin, 215
Tutti-men, 46–8
Twing-Twang, 221 *et seq.*
Tynwald Ceremony, 148–50

U

Ufton Court and Dole, 37–8
Up-Helly-aa, 10–14

V

Valentine's Day, 22–6
Verderer's Courts, 198–9
Vigil of St. Mark, 63
Vikings, 10–14

Vintners' Company, 151–2
Vision of Piers Plowman, 223

W

Wantage, 51
Wardmote of the Woodman of Arden, 159
Wassailing, 15
Wayfarer's Dole, 38–9
Well-Dressing, 80–4
Wells, Holy, 80–100
Westminster School, 19
Winifred, St., 94
Witches, 118
Wroth Silver, 211–12
Wykeham, 38

Y

Year, New, 1
Yeomen of the Guard, 55, 197